THE ART of
PODCASTING
on a
BUDGET

THE ART OF
PODCASTING
ON A BUDGET

amanda greenman

ISBN: 978-0-9991116-3-5

First published in the United States of America in September 2019.

Green Lane Press

Holland, Michigan

www.greenlanepress.com

Visit the author's website at www.amandagreenman.com

Cover and book design: A.J.G.

For my family, who always support me through my many, many interests, and to everyone who listened to our podcast for five years.

Contents

Chapter 1: Welcome to Podcasting

With organization, perseverance, attention to detail and a lot of enthusiasm, anyone can build a podcast that is as much of a pleasure to listen to as it is to create. It can all be accomplished without a professional set-up, with very little investment and no prior experience in podcasting. Welcome to the art of podcasting on a budget!

> *"Let's start a podcast!"*
> *"It will be fun... right?"*

In July 2013, two ladies who enjoyed talking for hours on end realized they wanted to talk about their passion with more people than just themselves. More specifically, they were both overly enthusiastic about a network television series, *Once Upon a Time*, which had captured their attention for its creative retelling of well-known fantasies and fairy tales with a modern spin.

These two women included me, Amanda, and my friend, Brittany. In the middle of that summer between two seasons of the television show, we spontaneously decided we should record our conversations, theories and reflections about the show and package them into a podcast.

The only problem?

We had no prior experience podcasting, and a very small budget. Plus, back in those days, podcasting was still a niche hobby: few people had heard of a podcast and even fewer listened to one regularly. It was highly likely we would develop more than a small audience of listeners of this niche television show. Still, we quickly latched onto the thrilling notion of being podcasters.

Fast forward to a few years later, and we had built a community around our podcast with a loyal base of thousands of listeners... all without a professional set-up, with a very minimal investment and without any prior experience in podcasting. What we discovered along the way was that with a little organization, perseverance, some attention to detail and a lot of enthusiasm, anyone can build an engaging podcast that is as much of a pleasure to listen to as it is to create.

In the years since, I have reflected upon our journey of building a podcast from scratch, with a very small budget, then cultivating and sustaining a base of loyal listeners.

I now know that a podcast can be an exhilarating, but also overwhelming, endeavour. It's more than just buying a microphone, recording a conversation and uploading it onto the internet. (Although those are important components as well.) Having a show required a lot of personal time and energy. We had to carefully tailor our passion and conversations into something that would be engaging for listeners, while remaining true to ourselves. We also needed to stay motivated and keep the project interesting for ourselves, so that we might be able to sustain the podcast long term. Early on when we first started podcasting, a fellow podcaster warned us that most podcasts didn't last beyond their first handful of episodes. We were determined not to let that happen to our show. Our perseverance and generally positive attitude, even when our show got off to a slow start and we had only about 14 listeners for the first several weeks, kept us going.

❂

Let me rewind for just a moment and set the stage for how our podcast was first born.

Once upon a time, I was slowly working my way through my master's degree thesis, a first draft of which was due in just a few months' time. I spent my days reading, researching, and working part time as a business English teacher. Oh, and procrastinating here and there. I was feeling a bit stuck in my head and wanted to have a hobby outside of reading and writing and, you know, working. In other words, I needed a better way to procrastinate.

Like anyone else might when needing a quick procrastination fix, I had found a few diversions on Netflix. My guilty pleasure at the time was watching *Once Upon a Time*. It was charming, had an intriguing enough story, and provided the perfect distraction whenever I needed it. (I should note that although I will mention this television show several times throughout this book - because our first and longest-running podcast was a fan podcast about it, after all - I have absolutely no affiliation with the television show, any person or entity involved with the series or network that created it).

I started chatting with an online friend about the show (that was Brittany). Fast forward a few months to the summer after the show's second season aired, when we were talking about it late online one night, dissecting different theories about how the storyline could go in the upcoming third season.

"Hey, we sound like podcasters," one of us said (I honestly

don't remember who it was).

Five minutes later, we were Googling "How to start a podcast," and a few hours after we'd read a dozen or so articles on the topic, our minds were made up. We were going to start podcasting about that television series, once the show's third season began airing at the end of the summer. We had just over a month to figure out how to get our podcast up and running.

Nearly five years later, on May 22, 2018, I uploaded the final podcast episode of our show, *Once Upon a Podcast*. Quite satisfyingly, it was episode number 150. Though I should note that, by then, we had uploaded even more than 150 podcast episodes. About a year and a half prior, we had started a second, completely different, podcast that had lasted 20 or so episodes. By May 2018, our podcast episodes for both shows had been listened to hundreds of thousands of times and we had built an audience of thousands of listeners from around the world.

To be fair, Once Upon a Podcast, and our short-lived second podcast, *Pop Culture Detectives*, were relatively small-time shows compared to some of the podcast blockbusters that have emerged in recent years. However, for a podcast that was originally intended to be nothing more than a hobby (and my personal procrastination tactic,) and something that we had hoped would maybe be listened to by a dozen or so listeners, in the end we produced show that had a larger audience than we had ever dared hope for in the beginning. It was enjoyed by a wide, diverse base of listeners and loved by us. We had learned and grown from the experience, and had a lot of good times (and a few trying times)… and, yes, we had done it all on a pretty tight budget.

Except for the few blog posts I read and rudimentary how-to videos that I had watched in the first summer we began podcasting (which had given me limited information about where to start and what to do, as podcasting was not nearly as big of a deal back then as it is now and there were limited instructions on how to actually make one), over the years we built the podcast from the ground-up, learning both the necessary technical and soft skills as we went along. Making, of course, plenty of mistakes along the way, but learning a lot from them.

Now, as I reflect upon my experience, over five years after I uploaded our first episode and nerve-wrackingly made it publicly available on iTunes, it seems as though podcasting is a totally different world, mainly because of its incredible leap in popularity. Nowadays, I don't have to explain what a podcast is to anyone, and most people listen to podcasts regularly. Every day, I notice that more influencers, celebrities, entrepreneurs and hobbyists are releasing their own podcasts. Many succeed in their attempts, others fail. I don't claim to know all of the secrets to podcasting success and achieving a podcasting "happily ever after." However, I do believe what I learned throughout my own podcasting process remains as relevant as it did in 2013. We really created our podcast ourselves, from self-hosting our podcast on its own website, which saved us from having to hire outside web designers, to editing our own show on free software, creating our graphics and logos, and even building our own intro and outro music. It was truly a "DIY" podcast on a shoestring budget. In this book, I hope to share how possible it is to start a podcast on your own, with a very small budget. Although there is no hard and fast rulebook about how to podcast, hopefully my experience will give you some ideas and context of what exactly goes into starting your own show.

Even more important than the technical aspects, I believe,

are the soft skills required to effectively podcast. In an era where only a few voices seem to be heard in mainstream media, it's more important than ever to find a way to share your message with others who may have similar interests or want to expand their knowledge and learn from you. I truly believe that everyone has a voice and everyone has something meaningful to give back to the world. Finding what that is is sometimes the easy part. So is setting up the equipment to record yourself or a website where you can upload your podcast. The part that takes dedication and commitment is knowing how to effectively spread your message to others, build a community of listeners, and make your voice heard.

With that in mind, in this book, I will focus on two equally important facets of podcasting. The first is the technical side of podcasting, including setting up the equipment you will need, recording and editing audio files, getting your files online and finally, distributing your podcast to as broad of an audience as possible.

The other aspect of this book is focused on how to shape, build, tailor and edit your content in a way that will help you gain listenership and appeal to the audience you wish to reach. Do you know what to podcast about? How can you tailor and edit your content to make it strong and compelling for your listeners? We will explore those questions and I will help you refine your plans for your podcast in a way that will enhance your chances of a successful, long-term show.

I also aim help you find your own podcasting voice, not in just a metaphorical sense but a literal sense as well. Having a strong, appealing on-air persona is an important aspect of podcasting. Have you ever sat through a college lecture or long business meeting during which the speaker drones on and on in the dullest possible way about a subject? In contrast, how does it feel to be sitting and watching a compelling talk or a stirring speech that might go on for an hour, but is so engrossing that you are left wanting even more? Some podcasters manage to leave listeners eager to hear more from them - that's a good thing, it means they will come back next week to listen to your next episode - and other podcasters only "inspire" listeners to hit pause on their app and search for another podcast. Our experience taught us how to keep the momentum and interest in our show strong, and I hope to help you accomplish that as well.

And of course, throughout this book, there is one other underlying key message I want to convey: that all of this can be accomplished on a budget. When I first set out to podcast, I wasn't sure it would be possible to produce a decent podcast without a professional recording studio, expensive microphones and audio mixers, and the best computers that money could buy. Indeed, most podcasters, at least back then, seemed to have been radio enthusiasts, music professionals or serious hobbyists before starting their podcast. My co-host and I wondered if we were the odd ones out, and if we could pull off a show with good enough quality on common consumer equipment and the laptops that we already owned.

To this day, from time to time I notice celebrities and influencers who are just starting podcasts posting images of themselves sitting in front of professional microphones in sterile recording studios, giving the impression that a large investment in time and resources, and access to the best possible recording environment, is needed before podcasting. Seeing photos of podcasters in what appear to be professional podcasting (or even radio or recording studio) setups can be discouraging if you are just starting out and are on a tight budget. This perception can lead to the notion that an "ordinary person" cannot join the ranks of a podcaster.

Well, we podcasted for five years on a reasonable budget. Over the years, several of our show's reviews on Apple's podcasting app even commended us on our audio quality. We always used microphones that cost under $100 each and I edited everything, without additional hardware like audio mixers, quickly on GarageBand, which is a free software that came with my MacBook Pro.

With this book, I hope to provide a start-to-finish guide-book on podcasting (literally start to finish, as I will share anecdotes from my own personal experiences from the early days of my podcast to the final episode release, and every-thing in between). I will also share some pitfalls and things to avoid as a podcaster to help you minimize the risk of fail-ing. Remember how I said that at one point I created a short-lived podcast of about 20 episodes? I had one successful and steady podcast for 5 years that started off all right and fin-ished very strong. But I also have the experience of having worked on a, for lack of better term, failed podcast. With this guide, I will also draw from that "failed" experience and share with you what happened, in the hopes that you can avoid having your podcast suffer from the same fate. Of course, I hardly consider that failure a completely lost cause, because I certainly learned a lot from the experience. I'm happy to share those "failed" experiences in this work as well, in the hopes that will help another podcaster avoid having to learn the same hard lessons we did.

But let's not dwell on failure at the moment. As the old adage goes, you can only really fail if you never try... so congratulations on taking the first step towards starting a podcast by reading this far. I hope this book will help you, a friend, anyone, understand how to not only create a podcast that is technically good enough to listen to, but also thor-oughly engaging and satisfying for both the audience and the creator over the long-term.

With that said, there is really no one clear, direct, set of rules that you have to follow to create a podcast. Indeed, there are more than one ways to successfully create a podcast. As such, this work is not a step-by-step instruction manual of every tiny detail that you will need to do or every rule that needs to be followed when creating a podcast. Instead, it is a series of guidelines, thoughts, ideas, and anecdotes that can provide you with a solid foundation for starting your own show. I want to help you get oriented and have a solid grasp of how to podcast. Yet, there is probably far more that will go into your creation of a podcast than can be fit into a single book, and there may be things that I never did that will be crucial for your podcast, and vice versa, some of what I say here may not apply to you or your show. What I do hope to accomplish is to provide you with a well-rounded point of view on what needs to, and ideally should, go into the creation of a podcast.

In a way, this work is the capstone of my podcasting career, an adventure that has been over five years in the making. I went from starting the podcast with a friend on a whim to truly loving the art form. What I like most about it is the fact that anyone can do it. Anyone can start a podcast, anyone can grow a community, and most importantly, anyone can share their voice with the world. I want to help others realize how easy it is to do.

When my cohost Brittany and I first set out to podcast, we thought we were pursuing a new little hobby that would take up a little of our spare time once or twice a week after work. What we ended up being shocked by were the conversations we provoked, the community spirit that formed around our show, and the truly global reach that our voices had. Our podcast allowed us to bond with some of the most unique people we had ever met, it allowed us to travel to new places, learn new skills and be exposed to a new industry. I hope that everyone who reads this book has the positive experience we had.

✸

Note from Amanda: Before I get too far into this book, I want to acknowledge that my podcast co-founder, co-producer and co-host, Brittany, was a huge part of my podcasting experience. Without her, I probably would have never had a podcast in the first place. I also credit a lot of our success in podcasting to her creativity and hard work. She contributed many things to the podcast that I could not have, and as such, I learned a lot from her. I was pleased when she agreed to write the following short piece to inspire you on your own podcast journey.

"If there's something you want to do, just do it!"

This is the sage advice you receive as soon as you lament to anyone, "I wish I could do _____." At least, it's the advice I receive as soon as I mention on social media that I wish I had time to travel more. "Just go do it, the only one holding you back is you!"

Yeah, thanks Susan, I'll get right on that, drop my job, use the last paycheck I got and go *Eat, Pray, Love* right to one of the most expensive places in the world; great advice, very solid.

Now, I'm going to tell you something. Bet you can't guess what it is.

Ready?

If there's something you want to do, just do it.

Of course, in the context of this book, I mean podcasting. When I started, even before I had my own podcast, I called into one I admired (more than once) because it meant I could talk about my favorite television show with someone who only wanted to talk about that show. I had no idea how to get in on it at the time, but sitting and chatting with a friend about a show we both love for as long as we wanted sounded like my version of heaven. Sometimes, navigating through people's Tumblr accounts and praying you'll connect through messages, and then hoping the messages turn into some mutual pining over pop culture is exhausting. I love reading other's well-written thoughts, but it isn't the same as having a back and forth conversation.

One evening, while Amanda and I were re-watching a previous season of *Once Upon a Time*, we were sending each other messages back and forth, talking about it, essentially 'live-blogging' but only to one another. At some point one of us (I can't remember which, but given my urgent need to tell the world all of my thoughts on television even though no one in the world asked, I'll take the blame) lamented it was a shame we couldn't talk about this on a broader scale to more people. Podcasts were only really becoming more popular thanks to Serial dropping its first season and becoming a massive hit, and there were a few other shows I listened to specifically about television; boiled down, even, to specific television shows.

"I wonder if we could do a podcast?"

Neither of us had ever had any podcasting experience that amounted to running our own. By this point I had done a few guest spots, participated in a few interviews with actors when a regular host was unavailable. Those things only involved showing up and talking from points already provided. So, for a while, there was a lot of "I really wish I knew how to do this" happening until finally that turned into knowing we both wanted to do it and just doing it.

From there, there was a mad rush of Googling what to buy, how to do various things (you haven't lived until you

Google "how to sound professional" - an idea quickly abandoned for authenticity) and figuring out how to gain listeners. In the beginning, Once Upon a Podcast was truly just for us. It didn't matter who was listening because we were getting to have a conversation "face-to-face" about something we both loved. It wasn't until people started reaching out to say "hey, I love your show and I want to talk to you about the thing that happened last week as well" that we realized "Oh. We need a Facebook page. Maybe some social media presence beyond Twitter."

You could probably read a hundred online articles that tell you how to gain a social media following, but the only thing anyone cares about, truly, is that you're passionate about what you're talking about for an hour or so. You could ask a hundred different engaging questions, but if you aren't engaged enough to interact or have an opinion or talk out a specific point with your listeners, then they won't care to only talk amongst themselves, not at first. You have to decide you want to build a community and from there, decide you want to take the time to cultivate it. You can make a page so followers can see when you post your latest episode, or you can ask questions, ponder them yourself, answer, respond, and become more than just a voice in an earbud.

None of this is platform specific and you may find that one is more your strength than another. [When publicizing our podcast on social media,] I handled Facebook and Amanda handled Twitter because it worked for us. Your audience will engage with you wherever you go. It may be small at first, but the one thing Amanda and I both held onto was that we didn't start a podcast for popularity or to even make a life-long career out of it. We just really liked a TV show and we valued one another's opinions.

So.

If you want to talk about a TV show, just do it. Google, read this book, reach out to other podcasts to interact. You never know where the journey may take you.

Brittany Duke, 2019

Chapter 2: Podcasting, a History

If you really hate history, you can skip this chapter... or not. If you love history and wonder how podcasts even started in the first place, then read on!

As a bit of a history nerd, I find it useful to take stock of how far we have come with a technology before figuring out how it's going to move forward. In that spirit, I wanted to officially start the content of this book with a brief overview on the history of podcasting. If you are already ready to dig into the hands-on aspects of podcasting, then by all means, please feel free to continue onwards to the next chapter. Or if you are simply triggered by the mere mention of "history" thanks to a not-so-nice 6th grade history teacher in your past, then please feel free to continue onwards. (Though I promise there will be no quiz at the end of this chapter. In fact, the questions I will be asking you can all be found in that next chapter, where I just sent all of the non-history fans! So, joke is on them, I suppose.)

Good. Now that only those of us who enjoy history are still here, let's get started. It was September of 2005. I was finishing my final year of university and had just purchased my first iPod, which was tucked into the pocket of my black jacket as I walked to class. I was playing MuggleCast, a podcast produced by a well-known Harry Potter fansite called Mugglenet, a site that I had checked almost daily for years back when the Harry Potter books were still being written and fans (myself included) spent hours debating and theorizing about how the series would unfold. The MuggleCast podcast, which last I checked is still active, had just launched a few weeks prior.

Podcasting was new in the early aughts. It was beginning to emerge with the wider consumer interest in iPods (although iPods were first released in 2001, sales of the revolutionary personal music device only took off in 2004, when the iPod stopped being compatible only with Macs and began to co-operate with Windows PCs).

The novelty of being able to pick and choose what to listen to on-demand as I was walking to class was not really a thrill to me, it was more of a relief. iPods felt like something that had been a long time coming. I was born in the mid-80s and am, as comedian Iliza Shlesinger refers to us as, "an Elder Millennial." I was just old enough to have been forced to listen to music on the clunky and impractical CD players in high school (they had already felt like a piece of outdated trash to my high school self, compared to the massive library of music and on-demand playing capabilities of the music software on my desktop computer at home). My Boomer parents had used record players and my Gen-X aunts and uncles had tolerated tape players and CD players in their youth, but growing up with an entire music library on my computer, I am pretty sure my patience would have run out if Steve Jobs had come up with the iPod a moment later than he did. In my last year of university, finally being able to own an iPod, which could hold most of my music collection, seemed like the natural progression of things. In short, in 2005, I was very pleased with my iPod, and to this day I remember it being one of my favourite technology purchases, overshadowed only by the iPhone that came a few years later.

Looking back at it, I am still surprised by how new podcasting was in the early 2000s. The first audio files had been distributed online via websites or web "jukeboxes" in the 1990s. I remember designing websites in the late 90s and used to embed brief audio clips into my webpages. A little embedded player with a play and pause button allowed a visitor to the page to start and stop the audio on-demand. With the widespread adoption of broadband Internet access in the early 00s, and faster internet speeds in turn encouraging the rise of infamous music sharing sites like Napster, large audio files could suddenly be easily shared amongst users.

RSS feeds were conceptualized in the late 90s by developers associated with Netscape. RSS feeds allowed audio files to be shared more automatically and efficiently. First used to deliver syndicated text-only news updates to browsers, then becoming popular in 2002 when The New York Times used a RSS feed to deliver news articles to interested readers, RSS feeds that could share audio files were key in the development of podcasting. Although RSS feeds were met with limited interest and barely noticed by users or developers at first, eventually, RSS feeds evolved to both aggregate and syndicate audio files in addition to text updates. A key figure in all of this was a software developer named Dave Wilner who built an "audio blog" capability for his colleague who wanted to post mp3 files of interviews he had done onto his blog using an RSS feed. This feature quickly caught on, and by 2004, audio files were being widely delivered via RSS feeds. Software then started to crop up that we could recognize today as podcast clients: early versions of the Podcasts app on Apple mobile devices, Google Play, and other similar podcast delivery apps.

The term "podcast" was first used in The Guardian in February 2004. In September of that year, there were 24 search results for "podcasts" on Google. By October of 2004, there were nearly 3,000 search results for "podcasts." At the time of this book's publication, there are over 2.04 billion results for "podcasts" on Google.

In June 2005, Apple added podcasting to its iTunes software and enabled a podcasting feature in its GarageBand music editing and creation software. That year, the first People's Choice Podcast Awards were held, and by December 2005, the New Oxford American Dictionary decided "podcast" was their word of the year and added it to its 2006 edition.

By 2013, when my cohost and I began to conceptualize our podcast, podcasting was well-known in internet circles. Surprisingly, however, we still had to explain what a podcast was to many of our family members and friends. I remember at the time, my husband jokingly said it was "a little geeky" that I was trying to podcast. Several times in our first year of podcasting, we laughed, self-deprecatingly, about how the first thing every podcaster gets asked at parties is, "Wait, what is a podcast?"

At a get together with friends later that year, I remember my friends being somewhat fascinated when I explained to them that I was podcasting, and yet really had nothing to contribute to a conversation about podcasts. They were politely interested in my unusual hobby, but had clearly never listened to a podcast, much less created one of their own, and so had nothing more to say to me on the topic. Needless to say, my co-host and I bonded over the awkwardness of explaining how we were podcasters.

Then, all of the sudden, something happened in October 2014. We were barely into our second year of podcasting when a little show called Serial was released. An investigative journalism podcast hosted by Sarah Koening and developed by NPR's This American Life, the series, which investigated a 1999 murder case, experienced staggering popularity and became a cultural phenomenon. Seemingly overnight, everyone was talking about Serial, *Saturday Night Live* even parodied it in one of their episodes that December, and all at once, "podcast" became a household word, with everyone and their grandma wanting to download Serial. In 2015, Season 1 of Serial had been downloaded over 68 million times.

You may not think that two ladies talking about a family-friendly fantasy tv series would have anything to do with an investigative journalism podcast about a murder investigation. But in late 2014 and early 2015, we saw a staggering leap in the amount of listeners downloading our podcast. We attributed that at the time to the astounding success of Serial, and more than that, for Serial making "podcast" a mainstream word. Although podcasts had been established nearly a decade earlier, they had been relegated to the domain of a niche interest. In my opinion, Serial popularized podcasting, and moved podcasts into the mainstream consciousness. We therefore benefitted greatly from it: as people became more aware of the sheer existence of podcasts, we found that more people sought out podcasts on topics that interested them, which drove more people to our show. More than once, we lightheartedly thanked Serial for our boost in success in that 2014-2015 podcasting season.

Today, the opportunities that podcasting presents to creators are exciting. On the one hand, there is still a reasonably low barrier to entry if one wants to be a podcaster. Just as we found back in 2013, all one has to do to start a podcast is, essentially, record an audio file, upload the file, submit the RSS link, show name and a bit of information about your podcast to a few popular podcasting apps. Promote your show and new episodes to attract potential listeners on social media. Lather, rinse, repeat.

The benefit that today's podcasters have that we did not have is that there are so many more listeners out there to attract to your podcast. Yes, there are many more podcasts, too. But consider this. In 2013, when we launched Once Upon a Podcast, there were around 250,000 unique podcasts. With that said, in that year, only about 12% of Americans had listened to a podcast (based on a study by the Pew Research Center). In fact, that number had remained pretty steady for years: four years prior, in 2009, 11% of Americans had listened to a podcast. In other words, when we started, interest in podcasts had been stagnant, with no signs that it would necessarily become more popular over time.

However, by 2016, awareness in podcasting had grown significantly. The number of Americans who had listened to a podcast had nearly doubled in three years; in 2016, 21% of Americans had listened to a podcast. By 2018, that percentage had risen to 26%. In short, today's podcast creators have a larger pool of listeners from which to grow their audience.

Podcasting is a medium that suits consumers' present interests and expectations. Podcasts are yet another form of on-demand media. Consumers, at this point, are not only accustomed to - but actually expect - media to be available to them, on-demand, any time they want it. Movies, television, and yes, podcasts, are all forms of media that people are eager to consume, but they only want to consume it on their schedule. Furthermore, users also like to curate their own unique media playlists. Creating your own playlist of podcasts selected from shows that are available on your own personal favourite topics is a luxury that we can all enjoy today.

Although today's podcasters have to compete with celebrities and influencers who, it sometimes feels, are starting new podcasts every day, I truly believe there is room for everyone in the podcasting space. For one thing, just as Serial did for our podcast, influencers and celebrities who continue to drive new consumers towards podcasts are constantly promoting the listenership of podcasts as a whole. In short, don't be discouraged that there are so many "big name" podcasters out there, because in my experience, there's room for everyone and the more that general audiences are in the habit of listening to podcasts, the better it is for everyone.

Today's environment for podcasters really contrasts to the environment we started podcasting in years ago. When we first started our podcast, the future of podcasting looked dubious. In fact, Apple had just dropped several podcasting editing features from GarageBand when we first started out, a discouraging sign, one that seemed to suggest that they did not believe in the longevity of podcasts. We discovered there seemed to be very little interest from media influencers and companies, like Apple, in helping podcasters evolve and grow as far as providing resources and support to podcasters. Although our podcast was very briefly (about a half a day) featured in iTunes' New and Noteworthy section of their Podcast landing page in late 2013, that feature drove almost no new listeners to our show. Sometimes we wondered if we had started a podcast right when podcasting was about to become an obsolete media form.

Clearly, times have changed, and they changed in the pod-caster's favor. And conveniently enough, nobody questions what a podcast is or asks why I created one anymore. In fact, throughout the past several months, I have received encour-agement from many different people in my life to share my experiences in podcasting, because they might want to start a podcast sometime, too.

Technology and media industries are also supporting pod-casts. Car manufacturers are making sure that the technology in cars syncs with podcasting apps, so drivers can download and listen to podcasts more easily on the go. Smart home de-vices by Amazon, Apple and Google allow listeners to play podcast shows via simple spoken commands. In 2018, Apple Watch finally added Apple's Podcasting app to the device, enabling users to listen to podcasts via their wrists, a symbol of Apple's renewed faith in the medium. Suffice to say, pod-casts aren't going anywhere anytime soon.

CHAPTER 3: DEVELOPING YOUR SHOW

"Knowledge emerges only through invention and re-invention, through the restless, impatient, continuing, hopeful inquiry human beings pursue in the world, with the world, and with each other." — *Paulo Freire*

One of the reasons I first believed I could successfully podcast on a budget without professional-grade equipment, without a recording studio and without expensive audio mixing hardware or editing software, was because I believed in the concept that my co-host and I had for our podcast. We believed in our ability to carry an informative, friendly, engaging and fun conversation that could attract listeners.

Our instincts on this were right. Creating a strong, focused show concept and fostering our conversation skills, on-air presence and dynamic with each other ended up being, in my opinion, the keys to our success. Not a microphone, or a recording studio, or a good website. A podcast creator can have the fanciest equipment and podcasting set-up in the world and thousands of online followers, but without a solid, engaging reason for listeners to actually listen to what you are saying, what is the point?

Therefore, before running out to buy microphones or setting up a home podcast recording studio, it is worth sitting down and truly thinking about what, exactly, you want to podcast about. So, welcome to the first hands-on chapter of this book! I hope to help you take some loose ideas for your podcast and form them into a more solid podcast concept. Even if you believe you have a good grasp on what your podcast will be about, this chapter may help you hone in on some things you are overlooking. So, grab your favourite beverage, a notepad and pencil and let's dig in.

Obviously, I cannot tell you how to decide what to podcast about. Only you know what you have to share with the world and what you are passionate enough about to devote a good amount of time and energy to. This is something you will have to determine yourself. What I can do, however, and what I will be doing now, is guiding you through a few questions that you should ask yourself before you go any further with your podcast plans. These questions are intended to help you clarify your vision and goals for the podcast. If you aren't feeling quite ready with a specific vision or goal yet, these questions should help steer you in the right direction. Conversely, if you believe you already have a solid concept in mind, these questions may help refine your existing ideas.

To save yourself a lot of time, confusion and frustration later, it is well worth doing this planning now. With that said, nothing you decide here and now is set in stone. In fact, how I would have answered these questions when I was first starting out versus how I would have answered these questions when I was two or three years into podcasting would have probably been fairly different. You can fully expect to evolve as a podcaster and, as such, evolve your skills and message over time. Some of my personal notes and experiences that I am offering to you after each question will help you see how my perspective grew over time, and may provide you with some insights into the type of process you can expect as a podcaster.

For now, ask yourself these questions and answer based on where you are right now. We all have to start somewhere; consider this your entry point.

QUESTION 1: WHY DO YOU WANT TO PODCAST?

Be honest with yourself: is it because you feel you have to, to promote yourself or your brand? Or perhaps it's for your job, to promote your company? Is it because you want to share your skills with others? Teach things to other people? Have an outlet for your excitement or enthusiasm over a particular hobby, television show or video game? Go ahead, write it down. No one is judging you...

When we first talked about podcasting, I was interested in podcasting on several levels. First of all, the opportunity to talk - and I really enjoyed talking - to someone about something I enjoyed, and recording it and then sharing it with the world, sounded really appealing. By 2013, I had been blogging for several years, and the static, isolating process of typing a bunch of words into Wordpress and hitting "publish" every week made me long for a slightly more personal connection with my audience. Being a blogger felt a bit like talking into a void. Although I knew that a podcast audience would, like blog readers, still be shrouded in a degree of mystery or anonymity, at least with my co-host, I'd be chatting with another human being in the process of making content, and throwing our interactions out into the world offered me a bit of variety from the blogging I had been doing.

Second, I had always enjoyed broadcast journalism. Way back in high school, I was fairly certain I wanted to pursue a career in journalism, possibly in television as a reporter. I dipped my toes into the water and began to build my reporting skills for a year at my high school in a broadcasting class. However, my career interests shifted when I chose to go to university that didn't have a broadcast journalism program. I dabbled in print journalism as a freelance writer for several years. Then, in my final year of undergraduate studies, podcasts emerged, along with a lot of online media and social media (Facebook was launched my last year of university). Part of me wonders why I didn't just start a podcast sooner, back in 2005, or another broadcasting-style endeavor like a YouTube channel, right at a time when I could have been one of the first people through the door. Hindsight is 20/20! At any rate, by the time I did get around to creating a podcast, I was drawing upon - and indulging in - my pre-existing interests in journalism and broadcasting.

That is just how I was drawn to podcasting, and, there is no need to actually have any previous interest in podcasting or knowledge of broadcast journalism before starting your own podcast. Whatever brings you to podcasting, no doubt any one of your existing skills, areas of interest or talents will come in handy, as a podcaster has to wear many hats and it takes all different types of people to make different and uniquely compelling content.

With all of that said, as the host of the podcast and the central figure that audiences will connect with, I think it's worth doing some introspective work and looking at why you are drawn to podcasting. Is it because you are excited to share your passion for a subject or topic with others? Perhaps you hope to use your podcast as a stepping stone to promote yourself as a professional or expert in your field and to further your career. Or maybe it's not about the topic of your podcast at all: maybe you want to get actual experience as a podcaster so you can add the skill to a resume and set yourself up for a related career. Maybe you want to learn more about podcasting in general so you can see if it is something that could help you find your voice or express your feelings and develop as an individual. Perhaps you just want to talk and have the satisfaction of having hundreds or thousands of listeners listen to you. There is no right or wrong answer here, and the answer is completely up to you.

Do not confuse this question with refining what your podcast topic should be (that's coming up next). Instead, allow this question to prompt some self-reflection. The bottom line is what do you, as an individual, want to get out of this experience? What excites you most about starting a podcast? Write that down right now, and put it in a safe place, somewhere you can find it later on. We will come back to it later in this book.

QUESTION 2: WHAT IS YOUR PODCAST GOING TO BE ABOUT?

Now that you have done some soul searching about what brought you to podcasting, it's time to get down to business and start refining your plans for the show itself. The first order of business is to select a topic for your podcast. Not only will this obviously drive how you refine the finer details of your show, but it is also a practical thing to figure out, as you will have to choose a specific category and keywords when you start submitting your podcast show to podcast distribution apps (like Apple's iTunes) in the near future. You will also need to eventually decide on images and music that will represent your show, social media profiles, and so forth. So, whittling down a category for your show as one of the first orders of business is a worthy thing to do.

Most likely, if you have already decided to start a podcast and have picked up this book, you probably already have a certain topic in mind. If you are still not sure about an exact topic, then my biggest piece of advice is to narrow down your idea to be as specific and clear as possible. With this in mind, there are a few things I suggest considering. First of all, I recommend being as specific and clear as possible when deciding on a topic for your podcast.

I learned this the hard way, through our failed podcast show, Pop Culture Detectives. Remember how I mentioned we had a podcast that we stopped doing after about 20 or so episodes? Well, a part of the show's failure and lack of long-term viability was that we weren't specific enough about what the podcast was supposed to be about. This was in stark contrast to Once Upon a Podcast, which was about a specific television show. It dawned on us too late that podcasting about a television show is a clear, precise concept for a podcast. Podcasting about the vague topic of "pop culture" was, at least in our case, far too general, overwhelming and cumbersome of a topic for the two of us to handle.

Therefore, when deciding what to podcast about, it is important to have a very clear topic. For example, creating a podcast about one specific television show and then putting out episodes of the podcast aligned to each individual episode of that show gave us a strong and clear focus for Once Upon a Podcast. The benefits were not only that it was nice to have such a clear and precise vision of what we were talking about each week, but we also found that our specific focus made it easier for new listeners to find us and become comfortable with what we were talking about.

Unless you are going into podcasting with a massive fan base of followers, it is likely that you will need to build most, if not all, of your audience when you first start podcasting. In our experience, if you podcast about a specific topic that people are interested in, most likely people will find you organically through searches for that topic or recommendations on a podcast app that will direct listeners of one podcast to other, similar podcasts on a similar topic. In our case, nearly all of our listeners found us by searching for the name of the television show we were podcasting about because they were watching that tv show and wanted to listen to other fans talk about it. When they searched the name of the television show, Once Upon a Time, potential listeners found a total of around 10 podcasts about that tv show. After clicking through and reading the descriptions of each podcast, they might try out an episode or two of a few, until narrowing down which podcasts (or podcasts) they liked the best. Those were the podcasts they would keep coming back to. Of course, a lot of our listeners found us other ways, whether it was recommendations from fans of the show, listening to another podcast about the same tv show and that podcaster mentioning our podcast once or twice, our guest "appearances" on other podcasts, and of course finding us through the all-important social media outlets, and so forth. But regardless of how they found us, more often or not, they found us because they were already interested in that television show and wanted to hear a podcast about it.

To better illustrate my point, let's go back to that second, "failed" podcast we created for a moment. When we conceived Pop Culture Detectives, we wanted to create a podcast about, well, pop culture. From our point of view, we really craved more variety and freedom than Once Upon a Podcast was giving us as creators. At the time, we felt that

Once Upon a Podcast was restricting us, since the podcast was so closely tied to the television series that it was about. My co-host and I actually have diverse interests and we like to consume a variety of media, such as movies, books, other television series, other podcasts, documentaries, comics... you name it, we were desperate to talk about a bigger variety of things that felt current and exciting to us each week. However, this newfound freedom in what we could talk about each week quickly, and ironically, turned into a very challenging and oddly constraining goal.

What is pop culture, anyways? Television? Movies? Books? Comic books? Celebrities? The reality was, pop culture is all of those things, and probably a lot more. In short, it was way too broad of a topic for us to cover each week. We were just two ladies who already had another podcast to run (Once Upon a Podcast was still going strong), plus we both had full time jobs and family and home responsibilities. We quickly discovered that doing a podcast on a broad topic like pop culture was going to take a lot of time to research (to sift through the news and find the most news-worthy pop culture topics to discuss each week). It also took time to watch television series or movies that we planned to discuss in each episode. Since we didn't have the time to devote to podcasting about individual episodes of many different television series, like we had with Once Upon a Podcast, often we would try to build an entire 45 minute podcast episode around an entire season of a show, which implied a tremendous amount of television-viewing time to "research" our topic of discussion, on top of the overall time it took to record and create the podcast episode itself. We quickly realized we were not "Entertainment Weekly" or a similar news outlet, where broadcasters devote their entire work-week to scouring pop culture news. Moreover, when we were recording the episodes, we realized the broad nature of the podcast did not allow us to deeply discuss the books, movies or television series to the extent we could with Once Upon a Time on our weekly Once Upon a Podcast, because Pop Culture Detectives just wasn't focused enough. Talking about an entire season of a television series in just a 30-40 minute episode only gave us time to discuss brief thoughts and impressions. In other episodes, we tried to talk about movies and books, but those were problematic too, for the same reasons. We quickly grew stressed and overwhelmed by the broad nature of the show, and simply found ourselves lacking the time, resources and energy to pour into a podcast with such a wide scope.

We realized that other pop culture podcasts that are successful (NPR's Pop Culture Happy Hour, for example), typically rely on a large group of hosts who can rotate week from week on the show, most of the personalities of those shows are more than just hobbyists and actually have full time jobs that allow themselves to be immersed in pop culture. In other words, we were two women trying to compete with full-time pop culture professionals. This show sank to the bottom of the search results, buried under 100s of other podcasts claiming vaguely to be about "pop culture," while professional podcasts like NPR's Pop Culture Happy Hour consistently remained on top of search results and more than adequately met the needs of listeners who wanted a weekly general pop culture fix.

So, the lesson here is, unless you are going to podcast full time (and maybe even if you are), I highly recommend being as specific as possible when deciding on a topic for your podcast. Yes, you may be pigeon-holing yourself a little bit. However, we found that having a focused podcast is actually far more manageable to set up and stick to long-term. If your podcast goes well, you can also be flexible with other elements, such as the format of the show and, if you wish, adding in personal stories, anecdotes and other aspects of your personality into the discussions on the show.

The other lesson here is be true to yourself, your abilities, time and other constraints. A dozen or so episodes into Pop Culture Detectives, we discovered we didn't have the capacity to take on a whole pop culture podcast ourselves. We were trying to do something that we could not do from a time standpoint, and on top of it, we were trying to be something that was already being done by a lot of others and by teams of professionals who could dedicate themselves to podcasting on pop culture as a part of their full-time job.

How can you narrow down your podcast topic? Well, think of it like a funnel or inverted triangle. You have the broadest possible topic on the top, and then gradually want to narrow down or eliminate themes within that topic until you arrive at the specific theme of your choice at the bottom.

For our podcast that was a success and manageable, the one about Once Upon a Time, we had narrowed our topic down pretty significantly.

ENTERTAINMENT > TELEVISION > TELEVISION FANTASY/DRAMA > ONCE UPON A TIME

(Pop Culture Detectives, on the other hand, would have simply fallen into "Entertainment." It was not narrowed down at all!)

Moving away from the topic of entertainment, a podcast about computer programming could be narrowed down to be more specifically focused on programming in Drupal. For instance,

Technology > Programming > Programming for the Web > Drupal.

Or, a podcast that is about fixing cars could be narrowed down to be only about Swedish cars, or perhaps only about Volvos.

DIY & Repairs > Automobiles & Trucks > Cars > European > Swedish > Volvo

Instead of making a podcast about classical music, make the podcast about being a cellist. Or instead of a podcast about travel, make it about traveling as a single woman on a budget. You could even make a podcast about yourself, but make sure it has a specific focus on who you are as a person: such as if you are a comedian, make it about your life on the road as a comedian; if you are a fashion designer, make it about your fashion advice for others; if you are a American living in Argentina, make it about your specific experiences and challenges as an expat in Latin America, and so forth.

You may also want to think about how evergreen the content is that you will be talking about on your podcast. By evergreen, I mean are your podcast episodes going to be very time-sensitive, like ours were (based on a specific television show that was on air at the same time), and possibly have an 'expiration date" or a date when fewer people are going to be interested in your show because it will be less relevant later on? News podcasts or podcasts about technology, something that changes quickly, are usually less evergreen. However, other types of podcast topics, such as a podcast on history, classical art, or culture, may be relevant for years, even decades, after you first put out your show, and are therefore quite evergreen. There are pros and cons to both types of topics as a podcaster, and I only have experience with putting out more time-sensitive content on a podcast. I have often wished that our podcast had had longer-term appeal. Again, there is no right or wrong type of podcast content, but it is good to be aware of what type of podcast your show will fall into, and whether the content of the episodes are likely to be appealing or of interest for a long time after the initial release date of an episode, or not.

Deciding on a specific topic may just be a launching point for your podcast. It's a way for you to find "your people" and connect with an audience on a topic you are passionate about. It does not mean that your podcast has to be 100% about that thing all of the time, and it does not mean that your podcast cannot change or evolve over time. Think of the specific topic of the podcast as the soil where you are putting down your roots. A carefully-selected topic will help you get through those first few podcast episodes by giving you direction as a podcaster and help you clearly show your audience what you are setting out to do. Once you achieve that and feel more comfortable and experienced as a podcaster, you can start to branch out, exploring how you can work other topics and interests into the show, switching up the format, having guests, and so on. But when you are just starting out, there are many benefits to honing in on one specific, focused topic.

QUESTION 3: WHO IS YOUR AUDIENCE?

Over the years, I have listened to many different podcasts. Many of them had a great concept and featured interesting topics. Sadly, however, I have found myself drifting away from most of the podcasts that I have tried out. Quite often, I feel that this is because the content creator does not have a good grasp on their listeners, or over time, loses sight of their audience and as a result can no longer connect well with them. As a listener, when I feel as though I can no longer connect with the hosts very well, I lose interest in the show and move on.

Our short-lived pop culture podcast was an example of this. Based on the metrics I tracked about our listeners, I noticed varying amounts of people would listen to different episodes depending on what the episode was about. This isn't necessarily a bad thing. But in our case, I believe it spoke to the fact that we didn't really know who our listeners were or what they were interested in. We would have dramatically different listeners (based on the feedback they sent us) depending on the show's topic that week. Again, that's not an unusual or necessarily a bad thing, especially if you have a large audience to begin with. But as a new podcast, I felt like we never were able to hit our groove and connect with our listeners in each episode. We never had a core group of loyal listeners sticking around and listening to most, or all, of our episodes on a consistent basis. I believe hosts should know how to connect with their audience well, week after week, and as a result, they will earn a base of loyal listeners and repeat listeners.

I have spoken with other podcasters who, when I ask them who their listeners are, have said vague things to me like, "our audience is anyone who likes xyz" or "anyone can be our audience." While this was not inaccurate - sure, anyone can subscribe to your show, and it's good to keep in mind that even someone who may not be at all who you thought would listen to your podcast might enjoy your show - in general, this was a poor approach to connecting with a core group of listeners. By not understanding their audience, who they were and what other interests they have, I felt these podcasters were unable to tailor their content well or find better, more effective ways to reach out and connect with their core group of listeners (or potential listeners). If we think about this in terms of marketing, most companies know who their target consumer is. Yoga apparel brands have a very good sense of who their target or typical consumer is. So do bath and body product shops, fast food restaurants, car companies, and so on. In short, if you have a good sense of who your target consumer or listener is, you can make more informed choices, especially when it comes to the content you discuss on your show, as well as where to publicize your show or how and where you try to connect with and reach those audiences.

Sometimes as creators we evolve and have to make changes to our content over time to reflect our own changing personal interests. That happens and is fine. Sometimes, you might lose some listeners when doing that. However, you can minimize the impact on your listeners if you know who they are and still strive to connect with them. The best way to keep an audience happy, we found, is to form a genuine connection with them. And to do that, you need to understand them on some level and know, essentially, who they are.

One thing that I think we did very well with Once Upon a Podcast was really thinking about who we wanted our audience to be before we even started our podcast, and then keeping that "ideal audience" very present in our minds throughout the life of the podcast. The easiest way to understand your audience is to think about them as real people. For a fun little project, go ahead and write down the names of some fictional characters that you think might like your podcast. Or, invent your own character who is an ideal audience member. What age are they? What do they do for work? Do they have family or kids? Do they have a career, or are they a student? What do they do for fun? Where do they live?

As your podcast attracts a larger audience, you may eventually get to know a few of your listeners and that will help you understand your audience even more. When we finally got to know some of our listeners, we were surprised and pleased to find that we had attracted many like minds. If you think about the type of people you want to connect with and work to create a show that you believe will resonate with them, you likely will eventually reach that audience you had in mind. With that said, we were also pleased to discover that we had some different types of listeners than we originally thought would listen to our podcast. For example, when we first started out, we thought we would really only attract female listeners who were roughly between the ages of 25-45. A few years into podcasting, we often received notes from two of our loyal listeners, a father and son who would listen to our show on their way to and from football practice. Not the audience we had in mind, but we were really pleased to have them among our listeners and it was a good reminder that you can never predict who is going to listen to your show!

Keep your target listeners in mind, especially when you are debating certain choices about your show or wondering if you are doing something that your audience will enjoy. This does not mean you have to completely cater to an audience's needs at all times. Nor should you think that all of your listeners are exactly the same. We had dramatically different people who listened to our show. We had listeners of all ages, backgrounds and who lived all over the world. Yet, I still felt like we had a good grasp of who most of them were and what most of them liked. When faced with a lot of choices, especially when you are first starting out with a new podcast, we always found it helpful to think back to our ideal or core audience member profile, and wonder if that person would appreciate whatever it was we were considering doing.

As podcast creators, we never really see our audience. We aren't on stage, with our listeners all sitting before our eyes. Podcasting can sometimes feel like shouting into a big, dark void. But, if you think about a few of the individuals who might potentially be out there listening to you, and keeping in mind they are real people with real interests and opinions, it may help from time to time, especially when you are trying to make some more difficult decisions about your show or feel as though you lack direction. In short, include your audience whenever you can!

As a bonus, here are a few more questions I recommend thinking about when it comes to your target podcast audience:

HOW OLD IS YOUR TARGET AUDIENCE?

For Once Upon a Podcast, we decided that our target audience was probably going to be the same age as the majority of the television show's viewers, somewhere between the ages of 18-35. However, since we wanted parents to feel ok about potentially listening to our podcast with their teenage children who might be as young as twelve or thirteen, we decided to not use any swear words that wouldn't be used on the television show itself (it was a primetime network show, so off-color language was pretty limited). The show itself regularly wavered somewhere between a "PG" and "PG-13" rating so we decided our podcast would roughly follow suit.

WHERE DOES YOUR TARGET AUDIENCE LIVE?

This mattered a little less to us since we knew we would naturally attract a lot of listeners in English-speaking countries, being an English language podcast, but we also knew that Once Upon a Time, the television series, was pretty widely available in various countries around the world thanks to Netflix, so we realized our podcast could have some international appeal. Over the years it was interesting to see where the vast majority of our audience came from, and there were some surprises. In the beginning, I assumed they would almost all be from the English speaking world. Our top three countries where listeners were from were indeed the USA, the United Kingdom and Canada. But the country with the fourth largest number of our listeners was Japan (we had nearly as many Japanese listeners as Canadian), followed by significant amounts of listeners in Germany, Australia, Sweden and Italy. A consideration of where our listeners lived did not have large implications on how we podcasted, as we were mostly talking about a television show and its fantasy world that was commonly known to all of our listeners, since they watched the show. But this awareness that not everyone listening was American (or even from the English-speaking world) did make a little difference in some of our casual conversations. For example, in November we would often talk about Thanksgiving, but usually our conversation might start out with "It's Thanksgiving here in the U.S.," rather than just assuming every single listener to know what day Thanksgiving fell on that year. That may seem like a minor detail, but I wanted to extend a genuine sense of inclusiveness for our listeners around the world.

WHAT POINT IN LIFE ARE MOST OF YOUR AUDIENCE MEMBERS IN? ARE THEY MAINLY STUDENTS, PARENTS, PROFESSIONALS, OR SENIOR CITIZENS, ETC.?

This consideration made a huge difference in how we approached Once Upon a Podcast. Considering the age range that most of our target listeners were, their 20s to mid-30s, we consciously decided to avoid talking too much about our significant others (since a lot of those younger adults who listened to us were not married), for example. We also tended to "curate" our casual "impromptu" conversations appropriately, often selecting topics to chitchat about that would be of interest to our listeners, trying to select anecdotes that would resonate with most. Although we were not obsessed with only catering to that type of audience, a mindfulness of the audience helped us to generally stay on track and helped us set the overall tone of our show.

IS YOUR TARGET AUDIENCE LIKE YOU OR ARE THEY VERY DIFFERENT?
HOW SO?

My co-host and I discussed this a lot. Since we were podcasting about a television series that we both enjoyed and when we started, we were in our late 20s - and therefore smack-dab in the middle of the target age range for the show - we were at an advantage in that we were pretty familiar with the target audience we were catering to, since we were basically one of them. Of course, a substantial portion of our listeners were still outside of that age range and wrote to us about their interests, many of which were different than ours. However, generally speaking, thinking about the similarities and differences between ourselves and our audience helped us figure out what we could connect with them most about.

With all of that said, I don't mean to suggest that you should fixate on a lot of stereotypes about yourself or your audience or make dramatic assumptions, but again, use it as a guide to help you make some decisions and better curate your content for your audience. This extends beyond age... it could be related to where your audience might live, their education level, their career or jobs, whether they have children, etc.

Does your audience want you to podcast about what you are podcasting about?

Keeping all of our previous questions in mind, it's important to ask whether your chosen topic and target audience mesh well. If not, where is the disconnect? Could you somehow reconcile those differences by making certain choices with your podcast?

Or, perhaps it doesn't matter. Rules were meant to be broken, after all. For example, maybe you are a former professional wrestler who wants to podcast about your three pet guinea pigs. A podcast like that might attract a wide range of individuals, from school-age guinea pig pet owners to wrestling fans... how could you appeal to both of those audiences? What choices would you need to make in terms of appealing to them?

QUESTION 4: HOW WILL YOU STRUCTURE YOUR PODCAST?

"How to structure your podcast" might make a good book in of itself. (Sequel, anyone?) By structure, I mean, how will you start out your episodes? How will you end them? Will you have a formal structure with different segments throughout your show? Will it be the same week to week? Will you have specific time frames for each segment, or a time limit for each episode? Or, will you have a totally unstructured show? There is no one right way to go about this. In our case, we listened to as many podcasts as we possibly could, then brainstormed about different ways to structure our show and how we could build our own podcast around some of the best practices others were doing.

It's good to note here that even if you want your podcast to sound and feel very casual, friendly and impromptu, generally speaking we found that the episodes always turned out better if we didn't totally "wing it." It is worth having some sort of plan or structure for each episode, even if it's just a brief outline with a few bullet-pointed ideas of topics you want to cover when recording the episode.

One thing that we found was that our audience enjoyed knowing what to expect during each of our episodes. Many of our listeners listened to our show as a way to relax on the way to work, or to wind down at night before bed. I think this is true of many podcast listeners: they tend to seek out podcasts because they are a way to relax and unwind, to shut off all of the visual clutter we are faced with each day and focus on the spoken word, much like storytellers who could soothe tired companions around a campfire after a long day. Therefore, a consistent structure in each of our episodes appealed to our listeners, much like a nice bedtime story has a familiar structure with a clear beginning, middle and end.

We found that sticking to the same structure in each episode was pleasing not only for our listeners, but also us, as it helped us quickly jot down notes before recording, as sometimes we lacked a lot of preparation time before we had to record each episode. (We used to put out our new episodes on a quick turnaround: we would watch the new television episode on Sunday nights and record our full podcast episode about that episode twenty-four hours later, on Monday nights. With our jobs and other commitments, we usually lacked time before recording an episode to write extensive notes or put a lot of thought into a new structure, so having a consistent format each week was a time-saving trick for us.)

Our episode structure was more or less the following:

- Opening music

- Brief introduction to the podcast episode, including what episode of Once Upon a Time we would be talking about and an informal chat about what we were up to that week in our personal lives - approximately 10-15 minutes

• In-depth discussion of the Once Upon a Time episode, formatted in a "Top 5" (prior to recording, we each separately prepared a list of our top 5 favourite aspects of that television episode and discussed them in a countdown format) - 30 minutes

• Wrap-up of our discussion, including our favourite "funny" moments that happened on the show that week and whether there was anything we were looking forward to in the next week of the series - 10 minutes

• Outro music

• (Sometimes I added a few bloopers or outtakes during our recording at the end of the episode, if I was feeling particularly ambitious when editing that week!)

Interestingly, although we put a lot of thought into our structure and were disciplined about sticking to it each week, for a few years we did not even know whether any of our listeners paid attention to or cared about the fact that we had a structure at all. That is, until, about three years into podcasting, and for a few various reasons, we felt the need to change up the structure a bit. (Early on in podcasting, we had a "news" segment where we discussed news about the television show, but eventually dropped that segment). This change resulted in more than a few letters and notes to us in protest! We had had no idea that change would be so alarming to some of our listeners. That taught me that our listeners were indeed comforted by the show's familiar format each week, and it is good that we had taken it seriously and really thought it out before starting our podcast. When we made those changes later on and dropped the news segment, and although the change was relatively minor, it was the only time we ever changed up our structure.

QUESTION 5: HOW FREQUENTLY WILL YOU RELEASE NEW EPISODES?

The importance of consistency will come up again in this book, but before we get to that, now is the time to start thinking about how often you want to release new episodes. Weekly? Monthly? Twice a month? Every other month? What makes sense to you? How much time can you commit to podcasting?

I found that, at a minimum, the recording process for a 45 minute podcast episode generally took 3 hours from set-up to putting things away. Editing the recording, exporting the recording into an mp4, and posting episodes online took another 3 hours. Promoting the episode after its release on social media, replying to questions and comments from listeners, and starting to plan for the next episode took at least another 3 hours. That's a 9 hour commitment per episode, minimum, and that was after we had a few weeks of experience podcasting.

Knowing that podcasting will be a big time commitment - and each episode is likely to take longer to research, record, edit, post and promote when you are first starting out - realistically, how much time can you spare for podcasting, considering your life, work and family commitments? I suggest setting a realistic plan for the frequency of your podcast episodes.

Once you set a plan for your episode release schedule, I recommend sticking to your schedule, as that will help you gain an audience. Not only do consistently-updated podcasts seem to gain more attention and more favourable outcomes in search engines, especially Apple's iTunes, your audience will also enjoy your dependability. Returning to the example of that outcry when our audience was unhappy about how we changed the format of our podcast episodes once in our 4 year stretch of podcasting, I can't imagine how upset they might have been if we were inconsistent with our release schedule every week. Consistently will help you build and keep an audience, because human beings love stability and predictability (at least when it comes to podcasting)!

Question 7: What do you hope to gain from podcasting?

This final, but critical, question sort of circles back to the first question. What do you want out of this podcasting experience?

Now that you have thought through these questions, do you have a clearer idea of what you hope to gain from having a podcast? You may want to ask yourself:

- Do you have any expectations?

- Do you hope to get paid?

- Do you expect to promote yourself or your business?

- Do you secretly dream of interviewing people you admire?

- Do you want thousands of listeners, or is a dozen ok?

- Where do you hope you (and your podcast) will be a year from now?

Again, there is no wrong answer, this is just for your benefit. You should write this answer down and keep it in mind.

It may also be interesting to revisit this question after, say, 6 months of podcasting, as what you hope to gain from the podcast could evolve over time.

At the end of the day, you are creating this podcast, your voice is a critical part of the podcast itself, this project is for you (and your co-hosts, if you have any). If you (and your co-hosts) don't love what you are doing and feel like you are getting something out of it (even if that something is personal satisfaction or the thrill of uploading your content to millions of potential listeners every week), you will be more likely to stick with the podcast over time.

When you have a clear vision of the type of show you want to create and put out into the world, as well as the type of listeners you hope to attract to your show, it should, first and foremost, keep you focused and motivated during what may be the more challenging times in your early podcasting days. If at any point you feel you've gone off track or if you feel like you have lost your voice at any time along the podcasting journey, instead of ditching the project altogether, revisiting why you wanted to start a show in the first place might one day help motivate you and re-focus your efforts when you are feeling less eager to move forward.

I suggest keeping all of your answers to the questions in this section in a safe place somewhere, as we will come back to a number of these subjects in future chapters.

CHAPTER 4: FINDING YOUR VOICE

How should you sound in your podcast? Do you even need to care how you sound on your show? Maybe you are hoping you will sound a certain way, perhaps more professional or friendlier or more likeable than you are in real life. Maybe you want to sound myste-rious and wise, like Gandalf or Obi-Wan Kenobi or Dumbledore. Is that even possible with a podcast? Is it recommended? Let's talk about that, young Padawan.

You have come up with a topic. You're now familiar with your target audience. Next, it's time to think about... well, you. The podcast host.

After all, the podcast host is the core, the heart and soul, of the show itself. As the creator of a podcast, not only will you be constructing the technical aspects of your podcast be-hind the scenes, but you will also be the face of the show, and that all-important voice that audiences will hear the moment they subscribe to your podcast.

As viewers of television, listeners of radio and, of course, podcast subscribers, we all have a good sense of how important a person - and their unique voice and personality - can be in truly defining a show. The host sets the tone, style and flair of each episode. Many famous media personalities are instantly recognizable not only because they are the host of a show, but because they have found a way to truly integrate their personality with the show itself, making it a critical part of a show's soul. Oprah, Jimmy Fallon, Whoopi Goldberg, Anderson Cooper, Rosie O'Donnell or Ellen DeGeneres: their own individual styles and personalities are what have truly defined the shows they were or are a part of and differentiate their work from others. They are all very unique people and that enabled their own shows to be unique, too. No one could confuse Rosie for Oprah or Anderson Cooper for Jimmy Fallon. Each of these individuals had a unique style or flair that has made their separate shows so recognizable and iconic.

Famous radio hosts have done the same. It's even more unusual, in some ways, for radio personalities to become well known, because they don't put their faces out there or have a constant visual presence to help them become recognizable. From loud, controversial and provocative political radio show hosts, to calmer, soothing voices on a local radio station's late-night programming, radio hosts convey a wide range of emotion, have distinguishing personalities, and even their unique voices are all key ways in which they have effectively connected with their audiences. From Delilah's soft, relaxed love song request show, to the fun enthusiasm that Ryan Seacrest has conveyed on his pop music-related shows, radio personalities have all cultivated their own unique personas that have differentiated themselves in a certain way and be memorable to listeners.

Podcasting is similar to both of these broadcast mediums.

As a podcast host, you will want to, both literally and meta-phorically, work on developing your own personal voice. How can you be true to yourself and best cultivate your personality to match the type of podcast you want to create? By considering the following core aspects of who you are and how you want to be heard, you will start the process of forming your voice and tailoring certain aspects of your personality that will help your work stand out.

Who are you?

When we were starting our podcast, Brittany and I were very occupied with trying to sound "professional." While there's something to be said for making sure to have some basic skills when it comes to speaking clearly and being relatively easy to listen to, what we should have been focusing on is the bigger picture of how to best bring out our most authentic selves on the podcast.

To more easily build a podcast and make sure it is true to you, I think it is helpful to take a moment to think about who you are and what makes you unique. Finding what makes you stand out should not be something that you need to fabricate: everyone has some quality or characteristic that makes themselves an individual. Recognizing what sets yourself apart from others, and especially other podcasters, and then cultivating these qualities, will help you begin to form your own unique presence as a podcaster.

In the previous chapter, I suggested that you ask yourself why you want to have a podcast and what you hope to gain from your podcast. These two questions should have helped you start to contemplate the type of show you want to create, who it would be for, and why you want to create that type of show and why you want that type of audience. For instance, do you want to have a podcast about your academic interests on 18th century history? Will it be listened to by students or researchers? What if what you are truly passionate about - beyond general history - is 18th century literature? Now how is your show - and audience - going to change? Is this topic a good match for who you are and what you are passionate about? Do you think you can connect with other like-minded individuals on this topic? If so, who are those individuals and why will they be drawn to your podcast?

As you can see, I have a lot of questions for you. In fact, let's take this a step further. To find out who you really are and what makes you tick and keeps you motivated, why not interview yourself? The goal of a little mini self-interview, if you will, is to help you dig into and understand more about why you are doing this podcast and what you hope others will get out of your podcast. In other words, let's figure out what is going to set you and your work apart.

The following questions that I suggest you ask yourself are a little different than those presented in the previous chapter because they are all about you, personally, designed to help you take stock of who you are as you go into this podcasting project. If you are planning on podcasting with a co-host, you may want to have your co-host answer these questions as well.

As a bonus, in the future, if anyone asks you why you are doing a podcast - whether it's a family member, friend, colleague, someone interviewing you or a member of your audience - you will have an answer ready for them. Trust me, this should come in handy at family barbecues or cocktail parties when someone inevitably begins quizzing you about your podcast.

I suggest grabbing a cup of coffee or tea, a pen and a notebook, and jotting down your honest responses to each of the following questions. As with the previous chapter, there are no right or wrong answers here. Don't be self-conscious if you answer some with what you think might be "wrong" answers, or by trying to be too perfect. Keep it real!

- Why do you want to podcast?

- Do you have a story to tell? What is it?

- What are you most passionate about with regards to the podcast topic you've selected?

- Conversely, what do you dislike most about the topic you selected? Is there any aspect of the topic that you want to avoid talking about on your show?

- What are you most passionate about in life?

- Is there anything you'd like to learn or know more about related to the topic you've chosen for your podcast? Do you want to research or learn these things over time, while you are podcasting, or do you think you need to know it all now? Why or why not?

- Is there anything you want to learn from doing this podcast?

- Is there anything you hope to learn from your listeners?

"I PODCAST, THEREFORE I AM": LET'S TALK ABOUT AUTHENTICITY

In an era of Internet personalities, from famous podcasters or YouTubers, individuals who grew up online and made their careers via social media platforms before they even graduated high school, it seems like it should be easy to put your life and interests out there and spin it into a career. Everyone seems to be able to post pages and pages of photos of themselves or breathtaking scenery, alongside tiny moody captions with their feelings. It's easy to believe that if you do these kinds of activities consistently, people will flock to your social media profiles and content, and you'll grow more and more popular, and people will love you and admire you... all just for being who you are.

I mean, it's easy to believe that. But the reality is, not all of us are going to be all that lucky at attracting a large audience with just a few posts here and there on social media.

First of all, though, it's always important to keep in mind that much of the internet is smoke and mirrors. In my experience, we all wear masks. Whether we're talking about how we all put on a "public" mask before going to work or school, or we're discussing an influencer who has made their career out of putting themselves out there yet only puts a small slice of their life online, we all typically have two sides: a public persona, that mask that is worn when we venture out into the world, and a private persona. Our private side tends to be our true selves. This is who we are when we shed that mask and are alone at home at the end of the long day, crashed on our couch and watching a movie, cuddled up with a fur baby and a glass of wine.

While we might have the best intentions about our ability to merge those two personalities, the public and private, and be truthful and honest and authentic about who we are when we are putting our work into the world, in reality, I believe most of us edit what we show to the world in some way. To not edit is terrifying.

Years ago, there was a skit on the cable comedy show Portlandia, about a couple who "cropped out the sadness" of their vacation photos on social media. The skit was about a couple who took a spur-of-the-moment trip to Italy. As viewers of the show, we saw the couple jet lagged, crashing in their cramped hotel room upon arrival, and sleeping through the long weekend before they realized they had misunderstood the time zone differences and, upon waking from their jet lag-induced sleep, had to immediately turn around and return on a long plane ride home without seeing anything beyond the inside of an airport and hotel room in Italy. Meanwhile, although they had only managed to take a few photos of themselves, they posted those few vacation photos on their social media profiles. When they returned home from Italy, their friends, missing all context about the reality of their trip and only seeing the few photos that the couple took on holiday, assumed they had had an amazing trip. The conclusion? The couple had "cropped out all of the sadness" of the few vacation photos they posted online, instead showing to the world that they had gone to Italy and allowed the world to assume they had the time of their lives.

In other words, in a world where everyone is famous and lives a public life, at least on social media, I believe we are all guilty of editing our lives a bit.

However, there is something a little different about the medium of podcasting. Maybe that's why we are drawn to it as podcasters, or listeners are drawn to podcasts instead of other types of media they could be consuming. In my experience, many people seem to be drawn to listening to podcasts because it's a little less glossy, a bit less edited, than so many other media types out there.

On the one hand, there's a certain kind of intimacy a podcaster has with each listener. I believe that listeners seek out podcasts for this. A podcast is a very immediate, unscripted (or less scripted), more impromptu style of communicating than videos, television, books or even social media photos often are. The more personal connection that podcasting seems to bring between a host speaking directly into the ears of a listener, on their phone, available anywhere and at any time, makes for a unique experience. A podcast's long format - usually podcast episodes are 40 minutes or so - allows the listener to have more time to get to know the podcast host in each episode. There's time for the host to dig deeper and share more detail, and there's time for listeners to relax into what the host is talking about.

In Once Upon a Podcast, we always felt like we had time to have a relaxed conversation about all of the talking points we wanted to include in our episode that week. The more relaxed pace of podcasting facilitated real moments between the two of us as co-hosts, fostering times of natural, spontaneous conversation.

For these reasons (and likely more), I found that listeners want a podcast host to be truthful, relatable and authentic. In fact, many of the most popular podcasts that I have listened to are ones where the hosts really try to connect with listeners in that "real" and relatable way and avoid sounding too stiff, scripted or otherwise disconnected. We had listeners who would tell us that they listened to us while falling asleep at night, as they were waking up during their morning exercise routine or commuting to work, caring for newborn babies, or on long road trips or long-haul flights. We shared more time with our audience than, say, an influencer who posts just one picture a day in a feed of millions of other influencers' photos, or a 10-minute video every week. As podcasters, we are with listeners during their everyday lives, during daily routines and relaxed moments. As podcasters, we are not Instagram celebrities or models who present a glossy - sometimes Photoshopped - snippet of our lives to their followers. On the contrary, a podcast host is a trusted - almost a surrogate - friend. We are their companion while doing mundane tasks such as driving, grocery shopping, working or falling asleep. Everyone's expectations of friendship may vary, but one common factor, I believe, is that there is something relatable, down-to-earth, and "normal" about the other person. This is where, as podcasters, it's good to remain as true to yourself and your life as possible in order to maintain that relatability with audiences.

Podcasters cannot distract our audiences with shiny photos or flashy special effects. The focus of our listeners is on our voices and the things we are saying (and how we are saying them). That is why taking time to think about the content of your show and how you will present that content during your show is a significant part of your work as a podcaster. Brittany and I spent a lot of time really talking about and considering how we wanted our podcast to sound to others and what we wanted everyone to get out of it. Even though we started out thinking we had to sound "professional", ultimately, what we discovered along the way, was that our audience responded best to us when we were true to our personalities and shared bits of our lives with our audience. Perhaps equally as importantly, being true to ourselves also made the podcast more sustainable for us: we had more fun, we were more relaxed, and it almost never felt like we were doing work.

I should mention here that not everyone would agree with me. We have all come across examples of how the Internet can reward people who are snarky, clever, sassy, even spiteful and mean. Viral videos, shocking "clickbait" articles, podcasters and radio hosts who constantly talk about extremely controversial topics or provocative subjects, or influencers with closets bursting with clothes and makeup in an effort to attract attention, do get popular. But in my experience, the amount of podcasts that offer interesting, informative, educational content far outweigh anything else available and in the long run are more successful. They may garner less attention, but day to day, when commuting to work or falling asleep at night, that's not the majority of what people want.

I say this because very often, the attention gained from "clickbait" methods will be short lived (it's hard to repeat shocking things on a regular basis, or if it is an ongoing theme of your podcast, at some point your audience will grow bored or grow immune to your shocking ways). In other words, I do not believe that faking your personality, making your content overly dramatic, or being extremely controversial is a sustainable plan. Instead, I have noticed that content that is informative, thoughtful, relatable, and reliable tends to attract loyal audiences and grow steadily over time. Such content creators might not get the short term thrill of having their work go viral, but they should at least be better positioned to enjoy the long-term satisfaction of building a meaningful, strong, loyal listener base over time.

So, this should be good news if you are not someone who is or likes to be shocking or outrageous or provocative. You can still podcast. No matter how tempting it might be to try something new for attention, I believe that by being true to who you are and what you can offer, you will be successful and enjoy steady growth in the long term. At the risk of sounding cliche, be true to yourself! And be patient as others find you, listen to your work, and learn over time that they can trust you and relate to you.

I believe that, with regards to those who are "fake" in some way, ultimately audiences will see through the charade of anyone who makes it a habit of putting up a front. Being obnoxious, loud, rude and shocking - or even just putting on airs or trying to be someone you are not - is going to grow tiresome and boring over time for both you and audiences. Audiences are also very good at sniffing out authenticity, and acting in a way that is not true to yourself will most likely ultimately become apparent to listeners and alienate them after a while.

This doesn't only apply to overly shocking or extreme content: an example of one common trap along these lines would be using vocabulary or phrases that are not natural for you to use in "real life". For instance, deciding to call your listeners "darlings" if "darling" is a word that you would never naturally use in your everyday life and have never called anyone else a darling before. It will probably not sound natural for you to say, and could sound very contrived if you use it as a nickname for your podcast listeners. (Of course, if "darlings" rolls off your tongue and you call everyone who you meet "darling" in person, then by all means go ahead and use it, in fact this might be the most affectionate and natural way for you to connect with your listeners!)

You may be starting to get nervous now, wondering how you will attract an audience if you can't invent a new personality, should not make up a quirky nickname or include "clickbait" topics in your podcast. I am fairly certain that every podcaster I have met, including myself and my co-host, have had the same kinds of thoughts: why would people listen to me? What if they find me boring?

(Of course, you can certainly do everything I am advising against, my point is that it may be difficult to sustain and might end up biting you in the back in the end.)

Keep in mind that sometimes reality is more interesting than fiction. Even though your own interests or life might seem mundane to you, they may very well be fascinating to others. By being yourself, the person you are comfortable being, with all of your nuances and quirks and unique personality traits, you will stand out from the rest, and also by keeping true to yourself, this will help you be more relatable to your listeners.

As humans, we connect to emotions.

An example of this was when Brittany and I were invited to interview some of the actors from the television series we were podcasting about at a fan event on stage. Prior to the event, Brittany was so overcome by emotion at the prospect of being asked to host such an event that she posted her thoughts and feelings about it online on her personal blog, a week or so before the event. To her surprise, one of the actors we were going to be interviewing actually saw the post and therefore knew who Brittany was the second we introduced ourselves to her in-person at the event.

Not everyone would put themselves out there like Brittany had that one time, but she was recognized and remembered for it, because of her genuine and open emotions. Consistently throughout our podcast, her willingness to share her genuine feelings were memorable and relatable, and that set her apart. She stood out for being herself.

When to edit (a little)...

Not to contradict the entire previous section (although that is literally what I am about to do), it is important as a podcaster to maybe edit yourself just a little bit. But before I get into it, get your pen back out! It's time for a few more questions that you should ask yourself in your self-interview:

> • What parts of your personal life might be of interest to podcast listeners, even if it's off topic? Do you mind sharing these details with your listeners?

• Is there anything you'd prefer not to share with your podcast listeners, for any reason at all?

• Which of the following would you be comfortable chatting about on your podcast and why? Is there anything that's a "no-go" and you'd never mention on your podcast, and why? Politics, religion, sex, relationships, children, family members, work, school, your health, the city/town/state/country where you live

• Would you feel comfortable talking about your family members in the podcast? Why or why not? Do you think talking about them could be relevant to your podcast topic or help you connect and relate to your listeners better?

• If you had to select one motto for yourself, or a quote that represents who you are, what would it be?

• If you had to pick a color to represent who you are, which color would it be and why?

• What are your greatest strengths as a person? How will those come through in your podcast?

• What are your weaknesses? How do you think those might change or interact with your podcast? Could revealing any of your weaknesses help you better relate to your audience, or would you rather not go into it or does it feel too off-topic?

What I mean by editing just a little bit is that it's ok to draw a line in the sand and decide on certain subjects that you never really want to divulge in your podcast. Some of the questions I listed above concern highly personal things, such as whether you would feel comfortable taking about family members on your podcast. (Or whether your family would be comfortable with you talking about them). You may also decide not to talk about these things for safety or security reasons.

While I believe it's important to be authentic and share about yourself and your life in order to connect with listeners, I also believe it's perfectly fine to edit what you share about your personal life, especially in cases where it will help you ensure your own personal comfort or safety, and that of your family.

Of course, I don't mean you should necessarily come off as closed or difficult to relate to, or podcast about high-risk topics that would require you to protect your family in some way. I am just suggesting that you should feel free to have some "no-go" topics on your podcast, such as not discussing your school-aged children because you do not want their names to be out in the open, or not talking about your political views because you are concerned about how it might impact your current employment. If you do ever get called out by listeners for omitting or leaving certain aspects of your life out, it's ok to let them know you have chosen not to talk about some topics for personal or private reasons. Your listeners should be understanding.

Furthermore, I think editing your conversation can help you streamline your podcast in general. By having a light hand in cutting out a few things you might otherwise talk about on your show, and by focusing your efforts into sharing things that you think your audience will be most interested in, this can help you avoid going too far down a tangent that might bore or otherwise confuse your listeners. While spontaneous and impromptu discussion may be fun and lead to some of the best conversations you will ever have with your co-hosts, on-air or otherwise, certain topics or themes or discussions may not suit your show and could distract from the overall listener enjoyment of your show.

We made a few conscious choices like this with our podcast. We intentionally eliminated a few topics from our roster of conversation topics. We actually felt that not talking about those things helped us connect better with our audience. In our case, we decided not to overtly discuss politics because we felt our listeners may be from a variety of backgrounds and have all different types of political points of view, and politics were totally irrelevant to the topic of our show anyways. We also made a conscious choice to not discuss our personal relationships or families in much detail because we felt that other podcasters who had shows similar to ours, on the same topic, were talking a lot about their families and relationships, and so we felt that by not talking about those things, that would help distinguish us from others who already had established podcasts.

If you are podcasting about a television show, it may not be necessary to ever mention details of your dating life or marriage on the podcast, so is that something you would ever bring up? If so, why or why not? Perhaps you would, because maybe your marriage is similar to the marriage of two of the characters on the show and you feel you can strongly relate and would like to share some anecdotal stories with your audience about your marriage. Or, perhaps you would not, because the show is a sci-fi show that has little to do with romantic relationships and you would rather keep your personal life private, anyways.

On the other hand, if you are podcasting about the actual topics of dating or marriage, discussing your experiences in relationships may be integral to your podcast episodes. If that is the case, try to plan out whether you need to set any more detailed boundaries about what you would or would not discuss on the podcast regarding your personal relationships, just to be clear with yourself about what you are going to be comfortable having "out there" online for anyone to listen to.

As an aside, and this probably goes without saying these days, but once something is online, yes, you can technically move the file or delete it, but I always consider in my mind that once something is online it can never really be erased. After all, people download files, make copies, and can generally do whatever they want with it once it is uploaded. So if you do make a "mistake" when recording your show and need to edit something out, be sure to do it before you upload the episode online. (And be careful during a live stream or broadcast, if you do those!)

Fleshing out how you want to approach certain topics on your podcast (if you want to even approach them in the first place) could make your life easier as a podcaster. In the long run, it will help you mentally sort out which areas of your life you want to bring into your show, and which areas you'd prefer to keep "behind closed doors" and to yourself. Remember that you are probably going to be speaking in an off-the-cuff, impromptu manner when recording episodes. Mentally figuring out beforehand what topics are no-go will help steer conversations in directions you are comfortable going.

There really are no rules with regards to how you decide what is okay vs. what is not okay to discuss in the podcast. Sticking to what makes you comfortable will also make you seem more comfortable and at ease in your recordings. I should also make it clear that these "rules" are meant to be broken, and that I do not feel like you need you need to stick to conversation topics that are directly related to your podcast; for example, maybe you have a podcast about murder mystery shows and you also have a pet parrot. Every week you mention your pet parrot and what she has been up to that week, and suddenly it becomes a "thing": stories about your parrot's life is an inseparable part of your personality as a podcast host. A pet parrot might not directly fit with the topic of the show, but the parrot may go over well with audiences and you may enjoy including it as a topic of general discussion. A memorable interest like that might even become something that helps you stand out and distinguishes your show from others.

For example, since Brittany and I were from different parts of the U.S., at some point we started to include a reoccurring conversation on our show in which we discussed our favourite regional expressions. She was from the southern U.S., and I grew up in the northern midwest, and over the course of a few weeks we had fun telling each other the expressions and names for foods and special holiday traditions that were typical in the regions we were from. This kind of conversation had nothing to do with the television show we were podcasting about, but our listeners seemed to enjoy it. It also ended up being a way for them to get to know us better. Many listeners even started writing in with expressions or particular words and traditions from their own regions in the world. We had listeners write in from Australia and New Zealand to chime in with their favorite regional expressions. As a side benefit, as hosts, this helped us get to know more about the different parts of the country and the world where our listeners were from. It helped us connect with and understand our listeners better.

A benefit to loosely planning and tailoring conversation topics is that it will help your show feel more consistent.

For example, if you always feature a small talk portion of your podcast and discuss topics like gardening, baking and your pet dog, you may find listeners look forward to hearing about those updates on your life and they will anticipate that portion in each new episode. Having reoccurring topics that you can always draw upon when you're not sure what to talk about during a small talk portion of the podcast can also help inspire your impromptu conversation during a recording session. Short on ideas for small talk that week? Well, your audience enjoys updates on what you have been baking.

Finally, never underestimate the value of having peace of mind. There's something to be said about authenticity, but there is also something to be said about the peace of mind you will have when drawing certain parameters around what you will and will not talk about on your show. Although my co-host and I were far from being famous podcasters, we did at one time have a listener who was a little too interested in writing into us each week. When we failed to acknowledge her letters and comments on a weekly basis, her attitude towards us grew hostile. She began to find out personal details about us through Google searches and wrote a plethora of negative reviews about us on podcasting websites and apps. At first, we tried to ignore her behaviour, but she was persistent and her messages and remarks resurfaced time and again over the course of nearly two years. Needless to say, I was glad that I had not talked about some of the details about my personal life in our podcast. I never talked about my family or personal relationships, never said the name of my spouse, and never talked about my day job or my spouse's job in detail. There was no easy way for listeners to find out about the more private aspects of my day to day life. Not to say it wouldn't have been impossible, but by being careful about what kind of identifying information you give, such as your exact residential location, birthdate, workplace and more, you can add a layer of protection and make it that much harder for any "overly interested" listeners to find private information about you.

CHEMISTRY LESSON

Many of Once Upon a Podcast's listener reviews contain something along the lines of how we as hosts had "great chemistry." Brittany and I spent a long time working on building our on-air dynamic. We very conscientiously wanted to create a fun and believable connection with each other on our show. Chemistry is a very intangible thing, yet humans can sense when two people have it. It does not have to be romantic, of course; chemistry could be two comedians who work well together on screen (Tina Fey and Amy Poehler, I am looking at you,) or a group of musicians who seem so much more dynamic and livelier when they are together on stage.

Ideally you will develop on-air chemistry with any co-host (or co-hosts) that you work with on your show. Listeners can recognize chemistry right away, and they will keep coming back to a podcast if they enjoy the dynamic and chemistry between the hosts. Our listeners frequently cited our on-air chemistry as one of the reasons they kept returning to our show. (In fact, by our last year of podcasting, some of our listeners were no longer even watching the television show that our podcast was about, and kept listening to our podcast because they enjoyed listening to us!)

There is no magic formula for building chemistry with another person, whether it is in real life or in a recorded podcast. If there was a clear formula for this, then dating would be a lost faster and easier, actors would always light up a theater stage together, and no story about romance or friendship in a movie would ever fall flat. (How disappointing is it when you read a book and there are two main characters with an incredible love story, only to have that book be adapted to screen and the actors fail to produce any of the same spark?)

Chemistry is elusive, but I think that on-air chemistry can be fostered. Podcasting with someone else is a bit like being in any kind of partnership or relationship. Whether you are getting married, going into business, or making a big purchase together with someone, it's important to have a good, solid, trusting relationship with them and overall enjoy being with that person. A strong relationship will help you both get through the good days and the bad, together.

One of the main ways we tried to foster our chemistry was by being truthful, honest and open with each other. In other words, we were vulnerable, and this translated to a very unique and honest type of dynamic both on and off air. The more you are willing to open up, be honest and communicate with your co-host on everything regarding the podcast, the more comfortable you will be with each other. The more you have a good, strong relationship behind the scenes, the more that will be apparent on-screen. Our working relationship was not always perfect: like any two people, we had occasional disagreements over the podcast, and from time to time one of us was in a bad mood. If your co-host is also a co-creator of your podcast with you, you both will need to make tricky behind-the-scenes decisions with each other from time to time. It's no different than running any type of business with someone else. Because we were committed to making our podcast partnership work (both on and off the air), and mainly because we were both honest with each other even when it was difficult to be honest, we were able to work out any disagreements and weather any bad days together.

I believe that our honesty and commitment to each other as co-hosts — and commitment to the podcast itself - really translated into our strong chemistry or dynamic together on-air. On her influential work about vulnerability, author Brené Brown has written, "Vulnerability sounds like truth and feels like courage. Truth and courage aren't always comfortable, but they're never weakness." In other words, honesty with ourselves and others results in vulnerability, and vulnerability empowers us to connect with others. As podcast co-hosts and co-creators, we were open and honest and truthful and sometimes even just brave when the times were tough and we were scared to share our real opinions or thoughts, but by overcoming any hesitations about doing so, we were stronger for it.

Another way we built a good working relationship was by embracing our differences as people. We did not have much in common outside of enjoying the challenge of podcasting and the television show we were podcasting about. Our podcast goals may have been aligned, but other than that, we lived in two very different places, had two different family situations, a similar work schedule but different careers, divergent interests outside of podcasting, and so forth. However, we are both open minded and were curious about each other's interests and activities. Often on our podcast, we played up our unique personalities. By using our differences to inspire a variety of on-air conversations with each other, and often riffing off of our unique personalities on air, we managed to imbue a wide variety of topics and unique dynamic into our conversations. I also think this helped more of our listeners connect with both of us. I believe that many of our individual listeners had more in common with one of us than the other, but once they listened to our strong dynamic with each other, in spite of our personal differences we tended to draw everyone into our conversations and help them buy into both of our personalities.

ALL THE WORLD'S A STAGE

There is one more key aspect to how I learned to find my voice and personality on-air. It also helped me cultivate a more interesting and dynamic connection with another human being. Very early on when podcasting, I enrolled in an acting class.

The class took place in a major city that had a strong tradition in training actors and entertainers, so I was lucky to have such a class available to me. Many of the other students in the class were aspiring actors, but several, like me, were taking the class to improve in their speaking and presentation skills, or to enhance their onstage presence as musicians or entertainers. At the time, I was the only one who was a podcaster and had an interest in working with online media, but I could imagine that now that may have changed. Several popular influencers have said that they have taken acting classes, and of course many actors have made the transition from their traditional work in television or movies to podcasting or other online media like YouTube channels.

It is not a requirement to take acting classes to start a podcast, but even though I was on a small budget at the time, I never regretted spending the money on taking the three month-long course. Not only did it make me more aware of how I could improve my presence in front of a camera (which is more useful than ever now with video-intensive ways to connect with an audience through platforms like Facebook, Instagram and YouTube), the acting class also helped me learn to use my voice to my best advantage in the podcast. In just a few weeks I learned how to control my voice better, in a way that helped me convey certain thoughts and emotions, and I was surprised to learn things like how physical posture can affect how I speak or sound. Beginner's acting classes involve active exercises and games, which serve to help you improve your overall presence, sense of space, as well as learn how to enhance and build a strong dynamic with others. Overall, I felt that understanding more about how to manage physical space, body movements, and tone of voice were very helpful as a podcaster. Taking that class opened me up to a whole new way of looking at communicating with others and fostering a strong and strategic presence on-air.

There is an art to using your body, your tone of voice, and all of your senses to create a persona and to better connect with audiences. Actors do this by digging in deep within themselves and using their toolkit of skills to bring a character (or even just themselves) to life in a way that is thoughtful, strategic and conveys a message.

As I mentioned earlier, I am all about maintaining authenticity as a podcaster. However, even if you really are being yourself in your podcast, understanding a bit about acting could help you bring out and enhance the aspects of who you are that you would like to share with your audience.

If you have the time and the means to do so, taking an acting lesson, even just a very introductory course, is an excellent exercise for anyone serious about podcasting. Community colleges and community theater groups may offer low-cost classes or workshops. I also imagine that similar types of classes or workshops on related topics such as improv comedy or public speaking would support the kinds of skills that a podcaster should develop.

CHAPTER 5: PODCASTING GEAR ON A BUDGET

Let's go shopping.

In 2016, I posted a video on YouTube about how to podcast on a budget. The video itself was focused on the equipment required as a podcaster. When I put the video online, I thought maybe one or two people would find it helpful if they were debating about how much to invest in a podcasting setup. However, the main oversight of that video, and generally the inspiration behind this book, was that podcasting successfully is more about having a strong podcasting concept and finding your voice (and working well with your co-hosts) than it is about the equipment. (This is also the reason that a guide to podcasting equipment is not the first chapter in my book, since buying the appropriate equipment for podcasting should not be your first thought when embarking upon your podcast adventure.)

With that said, to podcast, you probably will need a microphone and a few pieces of equipment and accessories.

As the title of this book suggests, I believe that the equipment required to podcast successfully does not need to be out of reach of anyone with a modest budget to spend on it. Everyone's idea of a modest budget varies, but I think that a podcaster can be up and running with a less than $200 initial investment (and with some cleverness and possibly help with friends who can lend you equipment, or with some secondhand bargain-hunting, it's possible to start podcasting with even less of an initial investment than that.)

I am a firm believer that you do not need a massive amount of fancy, expensive equipment to be a podcaster (or any kind of artist, for that matter). In the very consumerism-oriented world we live in, a lot of brands and stores are constantly bombarding us with an idealistic fantasy of who we could be if only we owned this or that new device or item or technology. Our lives would be better and richer if we just upgraded to insert amazing piece of equipment or technology item here. While there is some truth to the fact that "you get what you pay for" - in my experience, a nice microphone will sound better than the standard-issue microphone attached to your earbuds - I also believe in striking a balance between investing wisely on a few choice pieces to improve the quality, consistency and ease with which you podcast, while not completely blowing your budget with over-the-top equipment that will only minimally improve the quality of your work beyond a certain point.

THE ROOM WHERE IT HAPPENS

Before we dig into microphones and all of that fun stuff, indulge me for a moment as I ask you where, exactly, you are going to record this podcast? As I quickly discovered when starting out, where you set up your equipment in your home (or office) does impact the sound quality of your final recorded show. At the same time, you should not need to obsess over acoustics in your house so much that you have to go out and build an amazing state of the art recording booth. Before we started podcasting, when I was reading about how a home podcasting studio should be set up, for a while I was afraid I was going to fail as a podcaster before we even began for lack of adequate space. I didn't live in the right type of place to podcast. Everything I read about on the topic said that a podcaster has to have a dedicated, quiet, padded room to record in, much like a professional recording studio or sound booth.

I did not have any space in my home to even attempt to set up a small podcasting area. At the time, I lived in a small city apartment, and had no rooms (or even corners) to spare. I briefly considered converting my walk-in closet into a little recording booth, and even looked up tutorials online about how to do so (you basically pad your closet with blankets), but the fact that we were starting the podcast at the end of a hot August and there was zero air circulation in my closet meant that was just not going to happen. I did find some interesting tutorials for creating temporary or portable micro recording "studios" out of boxes and foam, and they did seem quite cute and cozy, but like the closet, you would have to figure out how to ventilate the space.

Years later, I still think a lot of people are obsessed with having the perfect podcasting studio space. Call me paranoid, but it almost seems like some people seem to want to create barriers to entry in podcasting by making it appear as though an elaborate podcasting studio set up is a necessity. Recently, it has become much more popular for everyone - including, of course, celebrities and social media influencers - to start their own podcasts. I have noticed many influencers on Instagram post "behind the scenes" photos of themselves in sleek studio-set ups. I have even watched a few videos by Instagram and YouTube stars about setting up their podcasting recording studio, then showing off a big, dedicated room with a sleek table, large professional recording microphones mounted to the ceiling, expensive headphones and even one or two employees sitting behind a mixer and laptop. I have never come across a photo by an influencer showing off an ordinary desk in an ordinary room with a $50 microphone in front of them... yet thousands of successful podcasts have been launched with such a setup.

Bottom line: anyone with a room and a desk or table to set their laptop and microphone on can have a podcast studio.

Once I nixed the idea of building a walk in closet-turned recording studio, the first question I asked myself was where, in my small apartment, could I realistically record and have the audio come out halfway decent (meaning, very little background noise and minimal to no echoes)?

I settled on a room in my apartment that had minimal noise and distractions at the time of day when I knew we'd be recording. This was a part of the house with no air conditioners or humidifiers running, far from the front windows where noisy traffic could sometimes be heard, relatively far from any of my pets inside or birds chirping outside. I also chose a room that had a lot of furniture and textiles in it, which resulted in better acoustics. Textiles deaden sound as it travels around a room and therefore help minimize echoing. Although echoes can be somewhat edited out later, I found that it was always nice to start with as "clean" of an audio file as you can by minimizing unnecessary sounds when recording.

Before I settled on my recording space, however, I tried a few test recordings in a few different places throughout my apartment, and I would recommend that anyone do the same.

Listen to how recordings of yourself talking sound different depending on where you are in your house, and even what part of a room you are in. This is an interesting experiment to do and will help you, through trial and error, achieve the type of sound you hope to have in your recording.

Although I would have liked to have podcasted at my desk, the room where my desk was set up - really just an area of our hallway - did not have enough soft furniture in it. As a result, my voice echoed a lot in that room. I didn't notice the echo in my everyday life, but it was very apparent in the test recordings I did. I realized this was because my desk where my computer was usually sitting on was large and had a smooth, finished wood surface, which bounced the sound around before being captured by the microphone. The floor was also a smooth, laminated wood, and at the time, I did not even have a rug on the floor in that space. The walls were decorated with a lot of hard glass picture frames and mirrors and I had modern, metal shelves. The curtains were thin, which let in a lot of light (a good thing because I live in the north), but those thin curtains did not do much in the way of absorbing sound. My voice sounded thin in the recordings in that room, almost like I was sitting in a tin can. In short, I couldn't podcast in my office.

The kitchen, another room in my house, would have been even worse, as the tile floors would have created way too many echoes.

I did try my closet (sans remodelling to make it a recording studio) and the acoustics were amazing in there. The clothes absorbed any errant sounds, so there was no echo thanks to all of the textiles, spare pillows, boxes and books in there. My voice seemed to go directly into the microphone, and it was strong and clear. The only problem? Well, as mentioned before, my personal comfort with regards to a lack of air circulation. Any fan I owned made noise, so I couldn't just set a fan in the closet with me. And the wifi signal also wasn't the strongest through the old, thick walls of the closet. However, doing a test recording in the closet did teach me about type of sound quality I hoped to achieve.

I finally settled on recording in my living room. Unlike the office area or kitchen, it had a huge rug over the wood floors, helping to absorb sound. I also love furniture, especially antique furniture, so my living room had lots of soft, upholstered wood furniture. Heavy velvet curtains at my window further helped absorb the sound and minimize the amount of smooth surfaces that could bounce my voice around. Plants and other "soft" decorations like throw pillows further helped control stray sound waves. The only part of the room that caused me trouble was the coffee table, which was made of glass. This is where I needed to place my computer and microphone when recording, and I noticed that the glass surface of the table was causing a bit of an issue with the sound quality, noticeably bouncing the sound around when I spoke into a recording. I tried putting a thick, fabric placemat under my computer and another one under my microphone. And you know what? It worked. The placemats seemed to immediately improve the sound quality and diminished the echo caused by the glass surface. A tablecloth, towel or desk pad on the coffee table would have also done the trick.

I will admit this setup wasn't perfect. One drawback of my living room is that in the summer, it is where my air conditioner is installed. When recording in the summer, I had to turn off the air conditioner because the noise was overpowering. A fan in the same room also made too much noise. Suffice to say, I experienced some very hot podcast recording sessions in summers. On the bright side, I lamented with other podcasters about this on Twitter one hot July night, and made a few fellow podcast host friends that way, as we all bonded over the misery of recording during a midsummer heatwave when air conditioners had to be turned off.

In the winter, my electric radiators in the living room sometimes made a kind of "clicking" noise while I recorded. From time to time I would edit that click out, as I could not turn off my heaters on a subzero day.

Another challenge was that my husband had to avoid that area of the house in the evenings when I was recording. If he walked around while we were recording, the creaking noises from the floor would definitely have caused noise in the audio. Since the living room was not a closed off space, there were definitely some drawbacks to not being able to close a door and record. Podcasters with children or pets wandering around the house should also be mindful of those factors.

With all of that said, you can really record your podcast anywhere. Unless you are going to invest in a professional studio set up and have the space and time to perfect your recording space, settling into the best possible option in your home or office - whether it is a kitchen, spare bedroom, living room, conference or break room in an office, or anywhere, really - is all you can do. I found that making a few small tweaks to reduce noise or enhance the acoustics went a long ways in helping boost the audio quality of your finished podcast. Tablecloths placed on smooth, hard surfaces, adding more curtains to windows, throwing a blanket or rug on a hard floor, and moving out hard metal or glass furniture from the immediate vicinity where you are recording, are all quick fixes to make a space have better acoustics. In short, I believe in working with what you have.

I believe that reaching a certain quality threshold is all that is necessary when it comes to podcasting. This means that your goal with creating an audio recording should be to have a recording that is comfortable for listeners to listen to and does not have a lot of awkward distractions (such as the hum of an air conditioner or the echoes of a voice in an empty room). In other words, your audio file should be clean enough to listen to that there are no major distractions that will take a listener out of what you are actually saying in the podcast.

Sure, if you had a professional podcasting studio or set-up, then your audio file might be flawless compared to a podcast recorded in a living room. However, I believe that all podcasters should focus more on meeting a minimum threshold for listener comfort with audio, rather than obsessing over creating the most perfect audio file ever. Listeners care more about what you have to say and the entertainment value of your show than sound quality. As long as they are not being irritated or distracted by truly poor, uncomfortable, and awkward sounds, listeners will likely forgive and quite possibly not even notice minor flaws. Aim to hit that "comfortable" quality threshold for listeners, rather than holding yourself to impossibly high standards and striving for perfection.

I also suggest not relying on post-recording editing to scrub an audio file clean of sound issues. Aiming to record the best, most comfortable-sounding audio file that you can manage to do is a more sustainable, less time-consuming plan than manually editing out distracting background noises from a file after the fact. Editing is time consuming and it is not always possible to remove every noise, echo or distraction in your sound file. Over the years, I also tried out a few filters that had supposedly been created for podcasters to "clean" an audio file of distractions, but never found they did much, if anything, to improve the overall quality of an audio recording. Manual editing (i.e. cropping out a distracting noise) can help you tweak the audio here and there when you miss something or an unexpected problem crops up, but setting yourself up for success from the start with a good recording space will help you in the long run.

Testing, 1, 2, 3... microphones

At the time of writing this book, my favorite podcasting microphone is still on the market: the Blue Snowball. This is a well-known favourite of podcasters, often making it into equipment lists for new podcasters and how-to blog posts on podcasting. I know many podcasters who use this mic. In fact, I don't recall exactly how or where I learned about this specific microphone before purchasing it, but I do remember that a lot of people were podcasting with it back then, and still are today.

I found that the main benefits of the Snowball include nice sound quality, a good price, relatively high level of portability and a very user-friendly design.

I found that my Snowball microphones (yes, microphones in the plural, I'll get to that in a moment) resulted in recordings with a clear, crisp sound quality that was clearly better than that produced with any typical earbud mic (such as the Apple iPhone earbuds) or headsets (such as the over-the-ear headphone and microphone headsets that are often used by gamers) on the market that either I have tried or that my co-host tried. We are not professionally trained audio technicians or sound engineers, so I cannot, unfortunately, get into the nitty gritty detail of why that was. All I know is that the microphone consistently created better sounding audio than the other budget-friendly consumer choices we tried. The audio recorded on the Blue Snowball always passed that all-important "comfortable to listen to" threshold we aspired to.

And did I mention it was budget-friendly? The Blue Snowball typically costs around $50 U.S.

The Blue Snowball is also small and self-supporting, so you don't need to mount it to anything, making it extremely portable. This also eliminates the need to create a dedicated space in your home or apartment for podcasting, since it can be set up right before you record and put away right afterwards. I always set it up on my desk in front of my computer when recording, and then put it away in my closet when we were done. If you live in a small space like I did, its ability to be stored is helpful. Its portability was also convenient because I often traveled with mine, so that we could stick to our podcasting schedule even when I was away from home on business trips or, yes, on vacation as well. Although it's not tiny, it always fit easily into a suitcase or backpack and was lightweight.

At such an affordable price in a convenient package, the Blue Snowball was not without its flaws. Unfortunately, both my co-host and I had to re-purchase our Snowball mics a couple of times over the years. We each had a Snowball that failed after about 2 years of use because the part where the USB cable plugged into the microphone wore out and no longer provided a secure connection. Collectively, we owned 5 microphones over the years: my cohost owned 2, with her re-purchased one also wearing out towards the end of our podcasting adventures. I owned three; my first one failed after a little over a year because of the cord connection issue, although admittedly I had dropped it a few times and carried it halfway around the world in a suitcase on a couple of trips, so I may have been partly to blame. My second, not the Blue Snowball but a similar model by the same company, called the Blue Nessie (now discontinued), also stopped cooperating after about 3 years of use (I was unable to figure out why it stopped working). I purchased a new Snowball just before the start of our final half-season of podcasting so that

I could finish our podcast run with nice quality audio, and I am happy to report that that last microphone is still working just fine.

I wonder if some of the flaws with the microphone have been fixed by the manufacturer over the years, since the device is still on the market. So if you purchase the microphone, you may not (or you may) have the same experience with your device. Despite its durability issues, I do think it is telling that the product is still on the market, and it certainly seems to be a tried-and-true accessory for many podcasters and other audio enthusiasts.

Before we both settled on the Blue Snowball, we had tested out a variety of microphones, including our Apple earbuds and different types of headset/microphone combinations. I believe that the Blue Snowball's reliability as far as audio quality was concerned, week after week, was the main reason the audio on our show was clear and easy to listen to, without static or a tin-like effect that is often apparent with some other types of "budget friendly" microphones that we had either tried ourselves or heard others use. We never had a complaint in five years of podcasting about our sound quality. No doubt that was partly thanks to the reliability of our microphones.

Over the years, we tried the Apple standard-issue earbuds, which were surprisingly, not the worst. We would not have wanted to record our entire 4 years' worth of podcast episodes with that as our microphone, as there was a clear difference in the sound quality when we played back recordings made on our plain old earbuds vs. the Snowball, but when we had guests on our podcast who did not own any kind of microphone other than those ubiquitous earbuds, the earbuds produced reasonable results. We also had to use them in a pinch once or twice, and the results were passable.

You may wonder whether to use your earbuds, a headset, or something else you have on hand, to avoid that upfront cost of a microphone. Once again, trial and error works here. If you're on the fence, my recommendation is to record a few test pieces of audio with a couple of different microphones. Play them all back, and see if you think you could listen to someone talk for 30 minutes to an hour with that audio quality.

SNAP, CRACKLE AND POP: POP FILTERS

Several months into podcasting, my curiosity was piqued about pop filters. I had heard about them, seen the round, flat discs made of some kind of mesh attached to microphones, but had not yet tried using one myself. After editing a good dozen or so podcast episodes, there were some audio flaws that I often tried to scrub from our files during editing (which is, of course, a time-consuming and a monotonous task). I wondered if preventing those audio flaws before they were ever recorded might be the answer to my editing woes, and a pop filter seemed like a potential solution.

Once I realized that pop filters cost less than $15 online, I went ahead and ordered one.

Some sounds that were starting to irritate me in our recordings (mostly sounds that I was to blame for, my co-host did an amazing job with her vocal elocution) included the noise your mouth can make when you have just a bit of saliva on your lips or teeth and open your mouth to speak, creating a light clicking noise. Also, in English, we have some breathy letters that can be uncomfortable to listen to, as they are pronounced in an airy way that might be undetectable when listening to someone talk in person but that breathiness can annoyingly be amplified in a recording. Words that start with p, b or h, can generate a puff of air that the microphone will pick up in more detail than your ears will when listening to a live conversation. Indeed, this breathy sound is where the name for the pop filter comes from, as a mouth will often make a breathy "pop" noise when pronouncing a word like "popcorn" or "puff." These were just some of the irritating sounds that I was starting to become frustrated with when editing our podcast recordings. I felt like they were a potential distraction for listeners, yet editing them out constantly was time consuming. I made a conscientious effort to stop myself from producing too many of these unpleasant noises when recording, and this awareness and effort helped a bit, but any other help I could get was welcome.

As soon as I got my pop filter, I noticed a major difference. While it didn't completely filter or remove every annoying noise my mouth could make, it did seem to cut down on a lot of superfluous noise, especially those breathy "pop" sounds.

There are a number of brands of pop filters to choose from when shopping online. I never pinpointed one brand that was extra good. I just selected one that had a good price and enough decent reviews. As soon as I attached it to my Snowball, I noticed an immediate improvement in the recording quality. A lot of those irritating noises were greatly reduced or eliminated in subsequent recordings.

A pop filter might not make an earth-shattering difference to listeners, as these little sounds were generally small details that many people might not consciously notice. However, as someone who has listened to many recordings over the years while editing, I did feel the filters improved the overall quality of the final episodes and generally contributed to minimizing distractions for listeners. It was a well-spent $15.

LISTEN UP: HEADPHONES

Headphones were a necessary part of podcasting for us. For one, we were not in the same room, or even the same country always, during our 4 years of podcasting. Recording a podcast was a fairly complicated affair, as we would listen to each other over Skype or FaceTime and record our podcast episodes much, like recording a phone conversation. Therefore, a pair of headphones was required.

It was also necessary for me to edit using headphones. I tried editing over my computer's speakers, but I found that speakers did not allow me to pick up on some of the smaller details or flaws in the audio quality that I could hear quite plainly through microphones. It was especially important for me to record using headphones since our listeners were probably usually listening to our episodes through headphones as well.

Over the years, I used a number of different types of headphones. I tried cheap headphones and expensive headphones. Wireless headphones and wired headphones. Around-the-ear, on-ear, earbud, etc.

Interestingly, I never found a pair of headphones that I preferred more than my basic Apple earbuds. And, again, since at the time, most of our listeners were probably using those as well, I stuck with them when editing. However, if you have a favourite pair of headphones or have a better quality pair, then by all means, feel free to use them. Once again, it may be worth experimenting with a few different headphones, if you have access to some different choices, to settle on your preferred type.

When using the earbuds while recording, be sure to check your computer's audio settings to make sure that the audio input is your external podcasting microphone (i.e., the Blue Snowball) and not the earbud's microphone (that simple mistake ruined a recording or two for us over the years).

COMPUTER & SOFTWARE

Tablet computers and smartphones that are capable of editing sound files, including podcasts, are available to consumers now more than they were when we started in 2013. So in the interest of full disclosure, I never tried - nor felt the need to - edit a podcast episode on my phone or iPad. Although I wouldn't rule out the possibility of using a smartphone or tablet to create a podcast from start to finish, as a podcast creator, I was much more comfortable using my laptop to edit my podcast episodes. Call me old fashioned, but I am actually much happier using a laptop to complete just about any kind of heavy editing task, whether it's an audio file, photo or video. Having my laptop was also far better when I had to create the website for our podcast and upload files to the Internet.

I used a MacBook Pro (actually, two different MacBook Pros) throughout the life of our podcast. I have been a loyal Apple user for well over a decade, but believe that a Windows laptop would also serve a podcaster well. My co-host had a MacBook Air for the majority of our podcasting life and then briefly used a Chromebook towards our final episodes. We had very poor success with the Chromebook, as the audio options and customizability were limited on the Chromebook and it did not seem compatible with her Snowball. That experience taught me one thing: that if you are going to be a podcaster, make sure that whatever device you are using to podcast, 1) is compatible with the model of microphone you have selected to podcast with (i.e. the Snowball) and 2) is compatible with your chosen editing software (i.e. Audacity or Garageband), if you are the one editing the episode. If you are creating your own website and self-hosting your podcast, I also recommend having access to a computer or laptop.

As an Apple user, I used Garageband as the software to edit our podcast and exported it as an mp3 file, which is the format of each episode's audio file that we uploaded online and distributed to listeners. Garageband is free for anyone who has macOS.

I have also tried out Audacity. Audacity is a free software program used by many podcasters that can be used on both macOS and Windows. Audacity has many of the same features as Garageband and is, arguably, built even better for podcast editing. I ultimately preferred the Garageband interface, likely because it was a more native experience on macOS, but I could have very easily edited our podcast and accomplished the same things I always did when editing using Audacity.

Many have asked me over the years how to edit an audio file. This obviously depends primarily on what software you use, and secondly, what your preferred workflow is. I ended up watching a lot of tutorials about Garageband that were created by podcasters. This helped me understand how others used that software to edit a file, how to export it properly into the correct format and quality settings. I also looked up tutorials from time to time when I had questions about a tiny detail or encountered a small problem with Garageband. Generally, Google had all of the answers when an unexpected problem arose during my editing process. I also let myself play around a lot and do a lot of trial and error before I settled on a workflow that worked best for me for editing our podcast.

I will go into a bit more detail about how I approached editing in an upcoming chapter.

External Hard Drives & Other Equipment

There comes a time in every content creator's life when you realize you need some extra space. Some laptops come with multiple Terabytes of storage, but no matter how much space I had available, it always seemed like extra storage was necessary. Every 6-12 months or so, to save space on my computer, I would transfer over our podcast files (i.e., all of the exported mp3 files as well as the most recent "raw" Garageband working files that I had been using) to an external hard drive. Ideally, I should have done this evenmore often as a general good habit of backing up my work.

Having backups of not just your podcast's audio files, but also the intro and outro music you use, the podcast's cover art (in varying sizes for different uses), the graphics and backup files for your podcast website (if you have one) and graphics for social media accounts, and so forth, all take up significant amounts of space on a device and therefore should be given a home on an external hard drive, preferably a drive that is dedicated solely for use for your podcast. An external hard drive is is not necessarily a top priority early on, but sooner or later when your device's storage space is starting to fill up, it is good to have one to transfer all of your files over and free up some space on your primary device. Of course, from time to time devices do completely fail, so having backups will give you peace of mind and help protect your investments in time, energy and money.

When you dive into the rabbit hole that is equipment that other podcasters may claim you will need in order to podcast, suggestions that you may encounter could include any number of accessories or other specialty items, such as adjustable microphone arms, webcams, etc. I would not recommend investing in all of these accessories and extras up front. I never found a need for a microphone arm, though I could see the appeal of being able to adjust your microphone, especially if you do have a dedicated podcasting space. Webcams, cameras and lighting set-ups are often used by those who enjoy doing live video podcast episodes or other types of livestreams to promote their work. However, if you already have a laptop to edit your podcast, chances are you already have a good webcam built into it and a sufficient set-up for live streaming until you discover whether that is something that you really will want or need to do as a podcaster.

Interestingly, during a period in our podcast's run, we found that live-streaming ourselves was a good marketing approach for our podcast and helped us build a community around our podcast by allowing us to connect "face to face" with our audience. We did these streams on Facebook Live, which worked using the Facebook app on our iPhones. I actually ended up buying a lighting setup from Amazon, which included two bright "umbrella" style lights that helped illuminate me while on camera, which was especially useful when doing Facebook Live at night. Later these lights became a useful tool when I started making more YouTube channels. Admittedly, these lights were an unnecessary splurge and certainly should not be the first thing anyone invests in when starting a podcast. I have seen many podcasters who actually livestream their podcast recording sessions (we were never brave enough to do this!) or do special episodes or livestreams for their most loyal supporters or listeners who sponsor their show in some way, such as on Patreon. So video may end up being a part of your work as a podcaster. However, it is certainly optional and may not be the first thing you want to work on setting up when starting out on a budget.

When you first begin your podcast, you never know what other equipment you may end up needing and what kinds of gadgets might become relevant to you down the road. I'd suggest going light on the accessories at first. Save your money for the items you might end up needing later on, once you hit a rhythm and once you start to see what kind of tactics you enjoy using to enhance the audio of your podcast, connect with your audience, or market your show.

CHAPTER 6: ALL ABOUT EDITING

You're not done yet when you hit "stop" after recording. Next up, you have to decide how to - and whether to even - edit your episode. This is a bigger decision than it seems, and can drastically impact the tone and feel of your podcast.

The process of editing a podcast episode is more than just finding a software application or app that you enjoy working with audio files on - although that's important, too - it also can have a lot of influence on the overall style and feel of your podcast.

There are some aspects that every podcast creator should consider before digging in and editing an episode. This is where it's a good idea to remember what you originally hoped to accomplish with your podcast. Do you want to have a casual podcast that listeners can enjoy on their way to and from work? Or do you want a polished, well-produced podcast with various segments interspersed with music, something that sounds more like an audio book or documentary? What about a podcast that lands somewhere in between?

The tone and feel you want to achieve with your podcast will dictate how much time you spend editing each episode. The more well-produced and polished you want your podcast to sound, the more time you may want to give yourself to edit each episode. This by extension might also impact how frequently you can release new podcast episodes to listeners, as it may not be realistic to heavily edit a podcast after recording it and turn it around quickly, especially if you are not podcasting as a full-time job.

When it comes to podcasting, more editing is not always a good thing. In fact, we found it could even be detrimental, or at the very least, unnecessary. As podcast creators, we always felt that more editing might make our listeners feel as though they were listening to something that was too formal. We wanted to keep the tone of our show casual, light and spontaneous (though still have some more structured elements). Think about the difference between a live morning show on television or radio, and a documentary film. A live morning talk show is casual, friendly, silly, sometimes there are mistakes. The reason people listen to live radio or watch a morning show is usually for the sense of company, the feeling of hanging out with friends. A documentary, on the other hand, is usually more serious, more thoughtful, with a lot of careful elements that have been edited in (or out) to form a cohesive narrative.

In my experience, podcasts very often serve the same role that people have - and still - turn to live radio for. That is to say, the listeners like to relate to the show's hosts and as they listen to the show, enjoy the general sense of being amongst friends.

If that is the style of podcast you are aiming for, then you might want to consider editing less, or editing with a lighter touch. This is the approach that we took with both of our podcasts, so I can share some personal experience and insight on this.

That's not to say that podcasting is not for those who wish to create highly thought-out, well produced, extremely informational and educational content. In fact, I have met podcast creators who use the medium as a way to share stories and document people and places. Journalists use the medium, as do documentary producers. Many well-known podcasts are carefully produced shows, with well-planned and even scripted episodes, and a team of individuals produces and edits every episode. Although having a large crew of editors and producers on hand doesn't hurt when taking on an ambitious project like an extremely thoughtfully made, documentary-style podcast, podcasters on a budget (in other words, those of us who do not have a team of editors and producers on call), will have to keep in mind how much time, energy and resources you can set aside to produce a more demanding podcast.

Editing with a light hand

When I started listening to podcasts in the early 2000s, I listened to them as a student, enjoying new episodes while walking through campus in between classes. Listening to a podcast was like having a portable group of friends with me who were there to hang out for a few minutes at a time when I was in need of a break or diversion. To my knowledge, in the early 2000s, there were not a lot of examples of professionally produced podcasts along the lines of Serial. Instead, I recall that at the time there were more podcasts that basically just felt like a group of friends had sat down around a microphone, recorded their conversation, and put it online.

That's not to say there were not podcasts that were more professionally produced, just that I remember the style back then being much more informal than it is on many of today's podcasts. From what I recall, podcasting as a medium started within a more casual framework.

My early enjoyment of the more casual, impromptu style of podcasting influenced our approach with our own podcast. I didn't worry about a lot of editing, sound effects, transitions or "bumper" music, scripted discussions, and so forth. While we wanted to create an overall general structure for our show and have a strong concept and values underlying our work, as mentioned in earlier chapters, we were also conscious of the fact that we needed to turn around our individual podcast episodes in just a few hours' time. Therefore, heavy editing was not a realistic plan for us, anyways. This was all right: we didn't think that two friends sitting around talking about television show episodes each week required a heavily produced podcast. Not that it couldn't have been a painstakingly produced and edited show, however: for years, my husband listened to a Game of Thrones podcast that was extremely well-edited, produced, written, and even musically scored with their own soundtrack, and it was really a marvellous show and testament to the wonderful results that can come from very thoughtfully and painstakingly edited podcasts. For us, however, I knew my time limits to turn around a final podcast file, and also was aware of the fact that two of us were doing our podcast by ourselves (we didn't have a large pool of editors to call upon for help). Based on both the style of show we wanted to create and the limitations of our resources, lightly edited episodes is what we aimed for.

Another alternative is to simply never edit your podcast. You could upload it directly online after recording, or only edit in some intro and outro music, without editing any of the recorded discussion. Many podcasters we have met over the years have told us that they never edit a podcast episode, they simply upload them. They have commented that this keeps the feel of their show very "real". Other podcasters have told me about their unique workflows for editing: for instance, one podcaster told me that as they were recording their show they would note down the time during the recording when they made a mistake so they could easily go back to just that part of the recording and edit the mistake out later. This prevented them from having to re-listen to the entire recording. I tried this approach, but personally was never able to take notes while staying focused on recording an episode, so I always listened to our podcast audio file fully and edited the entire thing after we were done recording. That just goes to show that everyone's workflow when it comes to podcasting can be very different.

HOSTS IN MULTIPLE LOCATIONS

My co-host and I were always (give or take) about 1500 miles apart when co-hosting our show. Remote podcasting was a way of life for us. At exactly 7pm every Monday night during our podcasting season, Brittany and I would come home from work and text each other to confirm we were ready to record. Monday nights were podcasting nights, since the television show was on Sunday nights and we wanted to have a new podcast episode released to our listeners by Tuesday mornings.

We recorded our podcast with the help of Skype and an add-in app for Skype. We would call each other on Skype and, using a 3rd party recording software that I had purchased (Ecamm's Call Recorder), record the call. Note that as Skype became less and less popular in the later years of our podcast and (we felt) there was some degradation to the quality of Skype calls, we began to use FaceTime to accomplish the same thing (Call Recorder also had an app for FaceTime). Usually 3rd party software like this costs a small one-time fee to download and use. I purchased Ecamm's Call Recorder, but other software is available.

Not all podcasters who record their podcast with remote cohosts used this method. In fact, I have seen many recommendations for each podcast host to individually record an audio file of themselves during the podcast recording session. This audio file would be made by each host's own computer's audio input feature, using an app like Quicktime. The final recording would only include that one host's voice and not pick up anyone else's voice during the conversation. The conversation itself would take place via FaceTime or Skype and only be heard by the host through their earbud. The result of this is separate audio files, each with one individual host's side of the conversation that would need to be edited together into one combined, final episode. The result would, in theory, be a higher quality recording of each individual voice, assuming everyone's computer settings were set to record high quality input, than a recording that was taken of a Skype or FaceTime call. Those separate audio files of everyone's voice would then need to be placed together in different tracks in editing software and, when played back, all of the audio files would be harmonious and flow as one conversation since all hosts were talking to each other in the same conversational flow.

Once the editor would have all of the separate audio files aligned with each other, each separate audio track of each individual person's recording could be edited individually; this would be nice, for instance, if one of the hosts coughed or sneezed while another person was talking. When you only have one audio file from the entire conversation (i.e., a single group recording made from a FaceTime call,) this is not possible. If someone sneezes and someone else is talking, there's no way to cut that sneeze out without also eliminating what the other person was saying. Another benefit of this separate method of recording conversations is that some people believe it results in a higher quality recording, since the voice recording does not have to travel over FaceTime or Skype first. Each person is recorded directly on their computer and their actual conversation with each other over FaceTime or Skype is not recorded. In other words, the quality of the internet connection would not impact the quality of the audio recording.

My co-host and I tried recording like this and we felt that it was just not worth the extra effort. Since there were only two of us, we were both careful to not sneeze or cough while the other person was talking (often muting ourselves if the other person was talking for a longer amount of time about something), and we quickly learned how to not interrupt or talk over each other during a recording. This eliminated most need to edit out extra noises. If we did make a mistake and talk over each other or accidentally cut the other person off, we learned to stop, pause and restart the conversation so that it could easily be edited later on. I also felt the danger of having multiple hosts recording multiple tracks was that there were a lot more chances for error. You would need everyone to be sure that their recording was working and that the audio input was working properly. We both felt stressed out by this. Finally, our internet connections were usually good and we rarely had problems with the recording quality.

When wanting to edit the podcast quickly, I found that having to merge separate audio files and edit each one separately was too cumbersome for our show. Therefore, I preferred to edit from one audio file, the audio recording of our

Skype conversation, not two individual files that we had separately made.

For us, this ended up being an issue about what we preferred to do. I made the choice to sacrifice some of the audio quality for convenience and ease during the editing process; however, as with everything else, the choice is personal and I highly recommend trying both methods before settling on one approach.

MY PODCAST EDITING ROUTINE

We would usually finish recording our episode on a Monday night at around 9:30 pm. (If we were really powering through things, we might finish earlier, but we used to do a lot of chitchatting before actually getting down to the business of recording). I would then brew a cup of tea, grab a small snack, and set myself up on my couch (which is not recommended, by the way, as it's really not good for your back to sit on a couch, hunched over your laptop for long periods of time...) to settle in for a few hours of podcast editing.

I first dropped the recording I made of our call on Call Recorder into a GarageBand file and started to edit. With Garageband, I had a "template" that I used to open up, re-save with the name of that week's podcast episode, and edit within. This "template" was a file that I created myself that already had all of my preferred settings pre-saved into it, as well as the intro and outro music already dropped into one of the tracks. Having that "template" to drop each new recording in each week saved me a few minutes of set-up time. Obviously, adjustments would have to be made each time depending on how long the recording was that week (for instance, I would have to adjust where the outro music was positioned in its track because the recording lengths would be different from one week to the next), but for the most part, this template saved me a few minutes of setup time each night. It's worth thinking about whether you can find any similar shortcuts in your own workflow, especially if you are podcasting with as quick of a turnaround as we were each week.

Once I pasted my audio recording of our episode into the file, I settled in and listened to the entire recording from start to finish.

Editing is a skill that I found improved simply with time and practice. Our first few podcast episodes took me hours (as in, 4 or more) to edit a single hour-long episode. I probably was over-editing to an extent, but for the most part, it was because I was still getting the hang of manipulating the audio file, figuring out how to smoothly cut out mistakes or distracting sounds without making it totally obvious to listeners that I had edited the audio file. In other words, I wanted to keep the conversation natural. I also had to learn to live with some mistakes or flubs we had made in the recording, because our conversation might have sounded stilted if I cut out too much from it during editing.

Taking several hours to edit each podcast episode was frustrating, but it did get faster once I had more practice. Actually, one of the first things that made editing faster was that our skill as podcast hosts also improved. The better you do when actually podcasting - the fewer mistakes that you make, the clearer you speak and articulate your ideas, the less you say annoying things that should be edited out because they might distract the listener, like "um" or "ahh," eliminating awkward laughter and long pauses while recording - all of these things improve with practice. When you and your co-hosts get better at your conversations when recording, editing becomes much faster when there are fewer flaws or distractions to edit out.

After editing our podcast episode, which usually included editing the audio recording itself as well as dropping in intro music, outro music, and "bumper" music (transitional music between segments, which we had less of later in our show's life), I would export the podcast to a final mp3 file that would then be uploaded online. Garageband had an export feature that rendered all of my editing into one final mp3 file. Note that I actually did not export my show at the highest setting. Compressing your file is fine; you do not want a file size that is too big, which will take listeners too long to download, especially on slower Internet connections, or, if they are using their data service to download episodes, a large podcast file can really drain their data allotment for the month. Most podcasting software like Garageband and Audacity will walk you through the export and compression process when converting your working file to the final mp3 file. It may be worth researching what audio size or quality to aim for, as this has changed since we first started podcasting. We just followed what most other podcasters recommended at the time in terms of quality and file size (our show was exported in 64kbps, for those wondering, as were many of the shows of our fellow podcasters at the time).

Once I had my final mp3 file, it was time to get the show on the road - or at the very least, online. I used to queue the blog/podcast post to go live the next morning. I uploaded the final audio file to our website, then I would log on to our Wordpress site create a new post for that episode. This post included the podcast episode's information, such as title, description, keywords, cover art. In Wordpress, it was possible to schedule each post, so I could set everything up the night before when I finished editing and uploading the mp3 file, and have it go live online at 10am the next morning. More on how we self-hosted our podcast is coming up in the next chapter.

My podcasting routine did not stop there: it was very important for us to promote our new episode on social media. My co-host and I shared these duties, splitting up social media platforms (she handled Facebook, I handled Twitter and Instagram), and the next day we posted information about our new episode, responded to comments from listeners, and generally kept an eye on how the episode was doing that day and over the course of the rest of the week.

A FINAL THOUGHT ON THE ART OF EDITING

You may be wondering, if you want your podcast to seem casual and off-the cuff, what should you be editing out?

One option is to edit no part of your podcast and keep it as true to the original conversation as possible.

However, as podcasters, we found that we did not like this approach. Even though we were going for the feel in our episodes of being as real and natural as possible, we also did not want any ditzy mistakes, audio flaws or an overuse of "um" distracting our listeners from the actual content of the conversations we were having during our recording. We wanted our listeners to be comfortable listening to our podcast and not be pulled out of the real content of our show by a few too many "ums," a dumb mistake, or a brief drop in audio quality during our recording, caused by something like a brief blip in our internet connection. Those were the types of things we edited out.

When aiming to have a natural sounding conversation, it was difficult to determine which actual mistakes we made during our conversation that we should edit out. For instance, we sometimes made a mistake about a character name or confused one aspect of the show's episode with another episode. Mistakes happen and sometimes they can be endearing and make you seem more human or relatable. Those are the types of mistakes we tried to leave in. However, obviously dumb mistakes or mistakes that just felt like an unnecessary distraction were edited out. Mistakes can be very distracting; distracting enough that the listener will forget what you were talking about in the first place and lose focus on your conversation. A mistake might also take you too far from the image you want to portray in your show, such as swearing when you decided to create a "clean" show. Or, if you severely mispronounce someone's name and you don't want to seem insensitive about others, it might be worth editing that out. Over time, I got a better sense of what mistakes to leave in to make us more relatable and not make the podcast episodes feel "too" edited, and what things to just toss because they would be annoying for listeners. There aren't really any rules, rather, it's a matter of practice, of keeping true to your original goals of your show, and going with your instincts.

Um...

One of our most frequently edited-out words when we first started podcasting was "um." This is one "mistake" that every podcaster should learn to avoid. "Um," especially when uttered too frequently, is annoying, distracting, takes away from what you are saying, and makes you seem less intelligent than you really are. One or two "ums" here and there are normal and a human thing to say, but when podcasters are first starting out, a combination of nerves and not really knowing how many times in their daily life they actually do say this word, can result in a lot of "um" in an episode. A few episodes in after we began, I told my co-host that we both had to stop saying um because it was taking too long to edit enough of them out to not be annoying (and audio files were starting to sound too "edited" which was ruining our aesthetic goals of the show). We made an effort to eliminate um and managed to avoid saying it as much as possible while recording. After a year, I was spending little to no time editing out um. By the end of our run as podcasters, we might have said "um" once in a while, but it was not enough to be distracting or even be worth editing out. Practice really does make perfect.

Um is not the only Cardinal Sin that a podcaster can make when it comes to annoying vocal fumbles in a conversation. Many other words, phrases, habits and even sounds, such as vocal ticks like smacking the tongue to the roof of your mouth (you may have no idea what I am talking about until you listen to a recording of yourself and realize you do it all of the time), can be very distracting. Taking long pauses, awkwardly laughing when you are unsure of something, or just making genuine mistakes like mis-pronouncing something, can also be annoying.

During the course of a normal conversation, since we are human, we naturally will make some mistakes. I'm definitely not advocating scrubbing each of your episodes clean of all mistakes or imperfections, especially if you are aiming to have a casual, conversational style for your podcast. Ultimately, I found it was just best for our listeners if I did a "sweep" of each episode after recording to remove the largest distractions and biggest errors. By cleaning an episode of any obvious mistakes, you don't necessarily need to also eliminate out all of the small mistakes, too. Some of the smaller mistakes might go unnoticed or even add to the natural feel of the show.

I think what is key is finding a style that works for yourself and for your listeners. Remember that you are your own worst critic and are probably going to be harder on yourself than listeners will be. As a perfectionist, this was a hard lesson to learn. The idea is to strike a balance between not editing too much, and editing just enough to remove the big distractions. Again, the goal is to reach a threshold where the conversation is comfortable to listen to and sounds natural, but it does not have to be perfect.

CHAPTER 7: FROM YOUR MIND TO ONLINE

So now you have a podcast episode sitting on your hard drive. Now what? How do you make that podcast episode available to millions of potential adoring fans worldwide?

Since I posted my video on YouTube called "How to Podcast on a Budget," the single most common question I get asked is how do you actually get a podcast episode that you recorded and edited online?

I have also gotten the sense that this is the topic that intimidates potential podcasters the most, and may even cause someone to shy away from creating a podcast in the first place, especially if they have no background in web design.

Getting a podcast online really only requires three things. One, a place to store your podcast audio files (such as the final mp3 files you recorded and edited) online. This could be any kind of web host that will provide you with the space to upload your file so they are accessible online. Secondly, you need a RSS feed link that links your podcast uploads with platforms where end users, your listeners, can search for and find your podcast. These platforms include Apple's Podcasts app, Google Play, Spotify, etc. You also need a way to give podcasting platforms information about your podcast and its episodes, such as its name, description, as well as the titles and descriptions and keywords of each episode you upload.

There are two ways to accomplish all of this: one, build a podcasting website yourself, and self-host your podcast and all of your episodes, or find a specialized podcast hosting service to do all of the legwork for you. Although building the infrastructure to host a podcast yourself may seem daunting, it is what we did, and it has had a lot of long-term benefits, including being the most cost-effective and flexible option. With that said, the alternative, finding a podcast host to help you get your podcast online, may eliminate some of the initial research and set-up involved in self-hosting, and certainly will require less of a learning curve.

I have two main goals with this chapter. One, I hope to demystify the process a little bit by sharing what I actually did to get my podcast online using what I believe to be one of the most robust, yet cost-effective, means possible. I'll explain what steps are needed to understand the technical, behind the scenes aspects of doing this. I ended up self-hosting our podcast, so I want to explain why I made the choice to get my podcast online this way, and why I still believe it is the best way to put a podcast online.

Secondly, I'll provide an overview of the other options available to you, such as subscribing to a service that will essentially take your audio file and store it and distribute it for you to podcasting platforms. These other options may change over time, and even six months or a year from now there may be other choices available to podcasters.

A SELF-HOSTED PODCAST

How I got our podcast out into the world isn't the be-all, end-all way of podcasting, but I was very happy with this choice for the entire four years we podcasted and even now, a year and a half after we uploaded our last podcast episode. Self-hosting your podcast "from scratch" as I did offers a lot of flexibility and customizability, and once it is set up and you are comfortable with how it works, it is pretty easy and user friendly to maintain. I will warn you: setting up your own podcast this way does involve some more heavy lifting at the beginning, but in the long run, I found that the podcast basically ran itself. I was always happy with this approach. Since I was self-hosted, my podcast was not tied up with a podcast hosting platform, and therefore I was not at the mercy of an outside provider and beholden to their platform or payment requirements in terms of managing my podcast or worrying about the long-term viability of that provider or platform.

I have to admit that going into podcasting, I had the advantage of having a decent amount of knowledge about how to build a website and blog, as well as a lot of knowledge on coding. Please note, however, that if you have zero knowledge or experience with this, that's okay! I still think you can still self-host a podcast. It's a lot less intimidating than you might think. You may, however, have to do a bit of catch-up for in the way of searching for tutorials or instructions more than I had to, especially with regards to building a website. The first place to start if you consider this option is looking into getting a web host and registering a domain name.

When we had the idea to podcast, I already had subscribed to a web host. This meant that I had purchased a bit of server space (or, if you are a complete novice to this, it means you own space where you can upload your website to or build a website on; note that this is not the same as an URL, which just directs visitors to that server space). In my case, for a few years I had my own personal website and a freelance business website, and I had it self-hosted because I didn't want to worry about a free website provider doing something to my website in the long run (anyone remember Geocities? Which no longer exists, since no free website provider is guaranteed to last forever). Our site is hosted on Dreamhost, and I had a good experience with them: for a flat fee once a year (I have always paid a little over $100 per year), I had a steady space on which to host my website. Even better, Dreamhost offers a feature called "one-click installs" which facilitates installing Wordpress, which I will get to in a second.

I should mention that if you get a web host like Dreamhost, this does not include the registration of a website address, or URL. If you want something like "www.yourpodcastname. com", you will need to register that separately. Assuming that your URL of choice is not taken already, registering it is fairly inexpensive and can be done pretty painlessly through a lot of web hosts, including Dreamhost. The cost of this is usually around $15.00 or so per year, depending on how unique or in high demand your choice of URL is. Again, the URL, for those who are complete novices to this, is a way for visitors to actually find and arrive at your website. But having a server space to store your website on, like Dreamhost, is also necessary.

As an aside, you may want to search for potential URLs for your podcast before you settle on a name for your podcast, as URL availability may impact your choice of podcast name.

With a web host and an URL registered, you will be set to self-host your show.

At this point, you're probably thinking that, hey, you are here to create a podcast, not a website. Well, yes. However, I got our podcast online via a website. I first installed a new website onto the web host that would be specifically dedicated to the podcast. After registering our podcast's URL, I installed Wordpress, via Dreamhost's One Click Install feature, to that URL.

Wordpress is an open source website creation tool. In other words, it's the content management system that enables you to create pages, write blog posts, and generally manage a website with words, pictures, and so forth. Years ago, Wordpress was primarily used to create blogs, but now it is useful for all types of websites.

For podcast creators, Wordpress is one option to have podcast episodes not only accessible to visitors who want to download or listen to podcast episodes from your website, but it also will make it easier for those podcast episodes to be fed to podcast platforms, where most people go to search for new podcast shows or episodes, such as Apple iTunes or Google Play.

However, to do this, podcast creators will need to install a Wordpress app. Once I logged into my new Wordpress website, I searched for and installed a third party app called Blubrry. This app, which runs via Wordpress, enables podcast creators to very easily upload, manage, and post podcast episodes to their website. The app also features tools to customize the settings of how your podcast appears in podcast directories like iTunes, making it very easy to change the title of your podcast, the description of your show, and more, in addition to customizing every individual podcast episode that you will end up uploading, including the individual episode's descriptions and keywords. As someone who had used Wordpress in the past to publish a text-based blog, I found that posting podcast episodes with Blubrry was a very similar process and very intuitive.

As a little aside, none of this podcasting setup will work

with Wordpress.com, which is a free online blogging platform. The free blogs they offer are not compatible with
Blubrry. The type of Wordpress installed on self-hosted
websites is the "Wordpress.org" variety. Dreamhost had a
"one click install" option for Wordpress, which facilitated
the installation of Wordpress to my web space. The whole
process was surprisingly painless.

This book is not a detailed tutorial on setting up a website,
so if you would like to learn more about how to get server
space and register a URL, install Wordpress, and then install Blubrry, there are a lot of tutorials online that can assist
and answer questions based on your experience level. The
Wordpress and Blubrry websites themselves have a lot of in-
structional resources and how-to guides.

If you accomplish these steps, congratulations! You have a
way of getting your podcast online.

But there is one last step to distribute your podcast to listeners via platforms like iTunes: you will need to tell those
platforms where to find your podcast and where to grab all
of your podcast's latest episodes. Those episodes are then fed
to listeners. (In other words, the platforms themselves to not
host your podcast, you host the podcast and the platforms act
more as a search engine to find and direct listeners to those
platforms).

This is where the Blubrry app also comes in handy. When
you set up your podcast using the Blubrry app in Wordpress,
in the customization settings, Blubrry will give you an URL.
The address is not the same as your website address: it's a
unique address to the feed of your podcast episodes (otherwise known as the RSS feed). This is a link that is specially
designed to communicate with platforms like iTunes. It is
the URL that you will need to submit to those platforms.
Once you have submitted your podcast to the platform and

they have added you to their site (it usually takes anywhere from 24 hours to a couple of days for your show to appear on that platform), that RSS link will always "ping" the directory as soon as a new podcast episode goes live. Along with this signal, information about the episode (such as the title of the podcast episode, description and keywords specific to the episode, etc.) is communicated, so seconds after a new episode goes live on your website, users can see that a new episode is available and download it and listen to it on their platform of choice.

The final technical thing that I set up was installing an FTP program to transfer the final mp3 episode files from my computer to my server space. Again, this is much less intimidating than it sounds. Dreamhost had detailed instructions on how to link up a web space to a third party FTP client (there are dozens, if not hundreds, of FTP programs out there: my personal favorite was Cyberduck, but there are many more available). Simply follow the instructions from your web host and/or the FTP client, and you'll have a direct link from your computer to your server space in moments. Once I was logged into my server from my FTP client, files hosted on my web server could be dragged and dropped and rearranged, much in the same way that files can be copied from one folder to another on your computer hard drive.

My workflow for posting an episode

To recap, once a podcast episode was created (recorded, edited, and exported to a mp3 file saved on my computer), my workflow each evening when I was ready to make my podcast episode live would look a bit like this:

> • Drop an mp3 file into my hosted web space (via my FTP client, Cyberduck) in the "Episodes" folder I had created on my online server space

- Use a web browser to log into my Wordpress site

- Once logged into Wordpress, create a "new post"

- Write a title for the post that would contain the new episode, a brief description of the podcast episode, enter in a few keywords, upload episode art, etc. This is the same process as writing a new blog post, to those who are familiar with blogging will be familiar with this workflow

- When making a new post, the Blubrry app adds some additional fields that are specific to podcast episodes. The first field asks for a link to the podcast episode (mp3 file). After I dropped the mp3 file of the podcast episode from my hard drive to my server space, I used the FTP client (Cyberduck) to copy the URL and then pasted the link into that field.

- Next, Blubrry asks for some additional episode-specific information. Usually this is tied to the types of fields that podcasting platforms or apps will want to know about each episode, and this might change over time as updates are made to either Blubrry or to platforms like Apple's Podcasts or iTunes, or Google Play. I usually had to paste in a description of the episode as well as keywords specific to the episode.

- Like a blog post, Wordpress would then ask me if I wanted to schedule the post (this is handy:

we liked to release our episodes every Tuesday morning at 10am Eastern, so I usually took advantage of this feature).

• I hit either "Schedule" or "Post" and the work was done: the new podcast episode was sent out into the world.

Interestingly, in case you do have any listeners who enjoy going to visit a website itself rather than using a podcast app or platform, by using this set-up (Wordpress + Blubrry), the podcast episode will also be playable from your website itself, and the mp3 file can even be downloaded from your website. I actually did not think that any of our listeners would ever go to our website to play a podcast episode or download it as a file, but to my surprise, a fair number of our listeners actually preferred to do this versus using a podcasting platform or app. This feature does make it handy for those who listen to podcasts covertly, such as at work, and their work computers may not have an app like iTunes installed.

THE BENEFITS OF SELF-HOSTING

I found that self-hosting was a flexible option for us. First of all, since I built the platform on which our podcast "lived" by myself, I don't really have to worry about it ever disappearing for any reason. As long as I continue to pay my web hosting bills annually, my podcast episodes (and their corresponding information and posts) will also always be there. I do rely on Wordpress as my content management system (CMS), which fortunately has been a fairly tried and true CMS, so I expect it to be supported for a long time. I also rely on the Blubrry plug-in to power the way in which my podcast is communicated or delivered to podcasting platforms, which seems to also continue to be supported and updated regularly. Blubrry is free to use, though some advanced features require a subscription. It should be noted that if Blubrry ever disappeared on me, my website and podcast episodes would not disappear, as they are tied with my web server. The Blubrry app just facilitated, or streamlined, the connection and communication of information about my podcast between the web server space where the podcast episodes "live" and the delivery of those episodes to podcast directories like iTunes.

Cost breakdown of self-hosting

I podcasted with a self-built podcasting setup. It took some time to set up. Allow yourself several days if you are experienced with running your own website or using a content management system like Wordpress; if not, give yourself at least a week or two to work through tutorials and familiarize yourself with the process before you actually expect your podcast to go live. However, thanks to the inherent user-friendliness of Wordpress and Blubrry, once I had my podcast website set up, I found it was very easy to use and maintain. These systems are also well supported by their developers and I almost never had any issues or problems in four years of podcasting. Although set-up may sound complicated, it's achievable even if you don't have a lot of experience building websites, and it should not be difficult to maintain once the infrastructure is set-up and you have uploaded your first podcast episode successfully and posted it on Wordpress/Blubrry.

Web Host: As mentioned earlier, our web host was Dreamhost, which cost approximately $115/year, for unlimited space and transfer capacity. In other words, myself, and our visitors, could upload or download episodes as frequently as they wanted at any time; in other words, if a million people wanted to download one of my podcast episodes, they could, and I would not be charged more for the heavy traffic.

Website CMS: My content management system for the website was Wordpress (installed thanks to Dreamhost's easy "one click installs", a feature that is included if you subscribe to their service). This was free, because Wordpress is an open source content management system.

URL: I first registered a web address, or URL, with Namecheap, but then just transferred it to Dreamhost, because it was easier to manage from one place/one account. (Renewing the URL cost approximately $15/year.) It is not mandatory to have an URL, but is helpful for listeners to find you more easily online. Some users still prefer to go to a website dedicated to a podcast and download episodes themselves or listen to the podcast from their browser, and an easy-to-remember URL might be helpful for some of your listeners.

FTP Client to help you transfer the podcast episode (mp3 file) from your computer to the internet (your web host): CyberDuck. I believe I bought the client up front for a donation of $20. An FTP client is not mandatory, as files can be transferred a number of ways; your web host may have an upload feature and most web browsers can usually transfer files, however I found that a client like this one makes it a little more intuitive to upload podcast episodes from your computer to the web, and arrange them once they are online and stored in on your server space.

Blubrry: this app or plug-in that can be easily installed to Wordpress runs off of Wordpress and is easy, fast and free to install. This helps manage the posting of podcast episodes to your Wordpress website, as well as helps feed episodes and episode titles, descriptions, etc., along with manage settings of your podcast, to directories or podcatchers like iTunes. I found this app had many nice features, and was kept up to date fairly well (once, when iTunes changed things around with regards to how they wanted podcasts and information on podcasts to appear in their directory, Blubrry adjusted their settings accordingly, so users like myself could easily understand how to provide iTunes the information they needed according to their updates.) I am not currently familiar with any alternative to Blubrry.

A final thought if you are new to building a website or online presence: it is important to remember that you can't learn everything. Watch and read different tutorials on Wordpress and Blubrry, decide what is most important to you, and prioritize mastering the core steps of uploading your podcast episode and feeding it to podcasting apps like iTunes. Understanding or thoroughly learning every possible setting is impossible for most mere mortals and people doing this as a hobby: don't drive yourself crazy thinking you have to be an expert Wordpress user or Blubrry pro to be a podcaster. Learn what you need to know to get started, and skip the rest or save learning it for a rainy day project later on.

Plug and Play (3rd party hosting) options for podcasters

As you can imagine, not everyone goes the route of setting up a website and self-hosting their own podcast. In fact, when researching for this book, I was surprised by just how many podcasting hosts there are available and how many podcasters do take advantage of them. There are a lot of services that will manage your podcast for you, so you do not need to worry about buying your own server space, setting up an RSS feed, or maintaining a website and installing appropriate podcasting plugins. These "Plug and Play" options for podcasters are fantastic for anyone who does not want to spend the time learning about creating a website and setting up RSS feeds and so forth, as they will be able to enjoy a much more user-friendly interface. Podcast hosts will essentially be "plug and play" for those who want to get a podcast up and running quickly.

While the general concept and process of self-hosting a podcast has largely remained the same for many years, the specific options available for podcast hosting services seem to change all of the time. Indeed, this is one of the first drawbacks of finding a service that specializes in hosting podcasts: convenience also means sacrificing some certainty. What is the best, most affordable, and most well-rounded option now, may not be the best option 6 or 12 months from now. And a podcast host may also not be around, or offer the same options or packages, a few years from now, making your podcast vulnerable to long-term support by whatever podcast host you decide to go with.

With that said, when researching this chapter, I was surprised by how many podcast hosts there are now compared to what was available to podcasters only two or three years ago. When we started podcasting, I seem to recall only seeing a small handful of major players offering this service. At the time, Libsyn was a popular choice - and indeed one of the only choices - for podcasters who did not attempt their own self-hosted setup like I did. Today, other popular podcast hosts include Podbean, Buzzsprout, Simplecast, Spreaker and more.

I cannot give a full review of different podcasting hosting options as I have never used any of them. I have known podcasters who have used these options, and as such, have gathered some general knowledge and information about them over the years to offer at least a general idea of the pros and cons of selecting a podcast host.

The first benefit is that they usually have a simple interface and painless distribution of your podcast to iTunes, Google Play, Spotify, and other major podcasting apps. Most podcast hosts feature convenient tools that will allow you to easily upload your podcast episodes to the internet, and they take care of the backend setup involving RSS feeds and feeding your show and episodes to podcasting apps. This could save you some legwork and time. In addition, some podcasting hosts are very simple and truly streamline the backend experience for podcasters. Some podcasting hosts even provide podcasters with a free website or landing page, so your visitors will still be able to visit a dedicated online site specific to your show.

However, the first con is the cost of these services. Even though there are some annual costs to consider when self-hosting your podcast, those are usually an annual, flat fee that cover unlimited uploads and downloads. With most podcast hosting services, however, there are caps to how much you can upload in a certain amount of time or based on each episode (so, for instance, you may only be allowed to have a certain length or size of podcast episode file each week, or only be allowed to take up a certain amount of bandwidth per month.) This can quickly become very frustrating if, for instance, you want to publish a lot of episodes frequently, you have longer episodes, or if all of the sudden your listenership increases rapidly and exceeds your host's upload/download allowances for the month.

An unlimited plan with a podcast host will quickly become expensive - I calculated that with an unlimited amount of space on Podbean, for instance, that would have run us around $350 per year, which is approximately three times the price of what we paid for our unlimited web server space from Dreamhost. Another podcast host I saw charged $99/

month for up to 150,000 downloads per month by listeners, which means that to host a podcast with them, it would cost nearly as much in only a month as it did to self-host our podcast for an entire year. With that said, a lot of podcasting hosts have very affordable base plans; Libsyn, which has been around for a while, has plans that start for just a few dollars. Just be mindful that if your podcast audience grows and listeners download a lot of your episodes every month, you may be prompted very soon to upgrade your plan in order to handle the added traffic and downloads.

Back to the benefits of a podcast host: analytics. Understanding who, what, where, when, why and how listeners are listening to your podcast is notoriously difficult. The reason for this is, quite simply, that your listeners are not listening to a podcast via one route or method (some may download it directly from your website, others may stream it from their smart home device, yet others might download it to their iPhone via the Apple Podcasts app), and therefore there is no one easy way to calculate the specific volume or demographics of your listeners. However, some podcasting hosts offer their own version of an analytics tool, which could be useful if you hope to gain some insight into your audience.

Yet, with podcast hosts there is a lot of uncertainty. There are so many more platforms, podcast hosting services, and options available now then there were 5 years ago. But as with all things in technology, it can be difficult to know which hosts are going to still be around in a year, 3 years, 5 years, from now. What if they are not, but your podcast is? Another issue with uncertainty and stability in podcasting is that they may change the price of their plans on you, they may take away certain features or change other features in a way that is detrimental to your podcast or your workflow for podcasting, and so forth. When you self-host, this is never an issue. But when you are paying someone else to essentially set up your podcast for you, you have less control over the long-term implications and situation.

Another con is the reliability of podcasting hosting services. Admittedly, this is a consideration that should also be taken into account when you are searching for a good web host: someone who provides reliable servers and has good uptime is essential whether or not you are self-hosting your podcast. Usually, searching online for reviews and experiences provides a decent idea as to which services are the most reliable.

Finally, be careful about podcast hosts who automatically insert ads into your episodes. If something sounds too good to be true, like a free podcast hosting service or a deeply discounted podcast host, it is quite possible that the discounted service might insert ads into your content. Not only does this distract from your work, you will once again lose control over the content you are putting out into the world.

SUBMITTING YOUR SHOW TO PODCAST PLATFORMS

The final step of getting your podcast online after you have selected a host - whether self-hosted and managed or selecting a podcasting service - is to make sure that your podcast has been submitted to as many podcasting platforms as possible. Popular podcasting platforms change all of the time, but some of the more tried-and-true ones include Apple iTunes/Apple Podcasts, Google Play/Google Podcasts, Spotify, Amazon Alexa, etc.

One of the first steps we took as new podcasters was to submit our RSS feed to Apple iTunes. The process was relatively simple and Apple has a dedicated page for podcasters to submit their podcast for review. Simply search for Apple Podcasts or iTunes Connect, and the Apple website will walk you through the process. (You will need an Apple account to sign in and submit your podcast.)

We began podcasting before Google Play began, but nowadays I would also submit to Google Play right away as well. This is the podcasting app that Android users will have access to. Simply search for Podcasts in Google Play Music and you will be walked through the process. (Again, you will need a Google account to do this).

Other podcasting platforms to look for and submit your show to would be Spotify and TuneIn.

A NOTE ON PODCASTING ON YOUTUBE

Every so often I will see someone start a podcast and only release their episodes via YouTube. In other words, they could use a static photo (not have video) and upload a full-length audio file to YouTube containing their podcast. Although I am sure some people look for audio-only files on YouTube, YouTube is not the most popular or conducive platform to listen to audio-only content. It is, after all, a video platform. Over the years, there have been many podcasters who simultaneously upload their episodes on YouTube and release them as regular podcasts via the usual distribution routes. This might be a different way to reach new audiences, though we never found doing that paid off for our show.

In my experience, very few to none of our podcast listeners ever listened to our content through our YouTube channel, despite the fact that I uploaded some key episodes on there (including one with special interviews). Another time, we recorded a "special edition" podcast and only put it on YouTube (to save time) and our listeners complained and requested it be added to our normal podcasting feed so they could listen to it as an audio-only file.

I believe that most podcast listeners prefer the audio file format, and prefer it to be uploaded in a way that is easy to download it to their phone or mobile device or to listen to it when they are not connected to the internet, such as during a commute to work. I don't think that only uploading and distributing an audio podcast on YouTube will necessarily be worth the trouble, though this could change in time.

Chapter 8: The Art of Consistency

The chapter in which you realize that finishing your very first podcast episode is just the beginning. Building a successful podcast means putting out new episodes on a regular basis, and your audience will (hopefully) want another episode again from you very soon.

Routines. Some of us love them, some of us hate them. As a writer, I happened to have chosen a career that does not have a lot of strict routine: for better or worse, every day is different, and I work with different people and have to complete different tasks all of the time.

Even though I love freedom and flexibility and have a high tolerance for the unexpected, I have to admit that there are times when a strict routine is helpful, perhaps even essential. Podcasting was one pursuit that really seemed worth having a set schedule for. Having a routine resulted in consistency... for both myself and our listeners.

People are by nature creatures of habit. Our listeners

seemed to be no exception, and being able to anticipate the release dates and times of our episodes helped ensure that they would remember to return to our podcast each week. Because we were consistent with our release dates and times, listeners remembered us more easily and made our podcast a part of their routines. In fact, I saw many comments over the years from our listeners mentioning that that they looked forward to their Tuesday morning commutes to work or Tuesday afternoon walks because they knew our newest podcast episode would be available to them on that day and time.

I should acknowledge that this rigorous release schedule was bolstered by the fact that we were podcasting about a weekly network television show. The show's regular schedule kept us on task week to week, and also made our podcast very time-sensitive for listeners, because they wanted to listen to a new podcast episode soon after they had watched that week's Once Upon a Time episode. Our listeners were often excited about the developments on the tv show, and so were eager to hear our reactions and discussions in our podcast episode that would follow. We didn't want to disappoint them by releasing an episode late.

We made sure that we were releasing our new podcast episodes no later than a day and a half after the television show aired. Releasing a new podcast episode about 36 hours after the new television episode made sense to us: it was just enough time after the show aired for us to personally be able to reflect on the episode, jot down a few notes and record our thoughts in our podcast, but not so much time had passed that our listeners would no longer be as excited about the show that week.

Regardless of the subject you are podcasting about, think about how time-sensitive the content you are discussing is going to be, and whether that is going to impact when and how you release your new podcast episodes. Because of the time-sensitive nature of the content we discussed, we really needed to podcast on a weekly basis for a couple of months at a time, while the show's season was on air.

The time sensitivity of the content in our podcast aside, I really liked our weekly format of releasing new podcast episodes. I believe that a week was just the right amount of time in between episodes. It wasn't so long that our listeners completely forgot about us, but it also wasn't so frequent that they grew tired of us or that we didn't have enough content to discuss. Importantly, it seemed to be just enough time to keep our show's momentum going, but we weren't forced to record episodes so close together that we burned out on podcasting entirely.

I also found a lot of personal benefits to having a regular podcasting schedule. Like our listeners, I liked the predictability of our schedule. Knowing that I had to record every Monday night for a couple of months during each of our podcast seasons made it so I scheduled my Mondays at work in such a way to avoid having to work late that night. I also knew that I wouldn't be able to commit to any social obligations or other activities on those nights. Once I made podcasting a part of my life, just knowing that it would be take place on Monday nights helped me adjust and move everything else in my life around to different days or nights. Although I was giving up my entire Monday night for the podcast for weeks on end to record and edit our episodes, in some ways it was nice having the majority of the heavy lifting of the week for the podcast done on that night. If we hadn't had a deadline or timeline like that to stick to, we may have had a more difficult time agreeing on a day or time to record each week, and then I may have procrastinated throughout the week before editing and releasing the episode. Having to go through a lot of agony to schedule in different recording times or release dates for podcast episodes could have been a major source of stress and may have been time-consuming. Deciding once on a schedule and sticking to it forever was in many ways much easier than having no regular schedule.

With Once Upon a Podcast, because we had a time-sensitive podcast, I found that our listeners' habits were different than they might have been with other podcasts. They wanted to have a new podcast episode from us every week to correspond to a new episode of the television show. With our short-lived Pop Culture Detectives podcast, on the other hand, we released episodes on topics that were current, but not necessarily things that were so time sensitive that they had to be listened to the moment we released the new podcast episode. We also took advantage of this and released episodes less frequently; we decided to only release 2 episodes per month. Our topics were not as time sensitive. For example, we'd talk about entire seasons of television shows that had been recently released on streaming sites, movies that were recent releases, books and other related topics. So while the content was timely, it was not necessarily urgent to get out into the world.

This less-frequent podcasting schedule did have an impact on our listeners' habits. Tracking our downloads for Pop Culture Detectives over time, I noticed that our listeners tended to listen to our podcast episodes on an irregular basis. So, a newly released episode may have hardly any listeners for its few weeks, and then suddenly increase in popularity. Unfortunately, I believe our Pop Culture Detectives audience never got into the habit of listening to us every other week or checking for new episodes in that interval. It's possible that instead they would check out the feed every month or every other month, and then download all available episodes at once and listen to them at once. Or, they just took a look at the fed and only downloaded and listened to the topics that they were interested in, skipping the rest. Looking at our metrics on the show, I always noticed that some episodes had a respectable amount of listeners, others had hardly any listeners. In short, listeners for that podcast seemed to prefer to access our episodes on their own schedule, at their leisure, never holding their breath waiting for a new episode to drop every week. Perhaps that was because of the nature of the less time-sensitive content of that show, or maybe it was because we never really found a solid, consistent, loyal listener base for that podcast (or maybe a bit of both). It was very different than our listeners on Once Upon a Podcast, who were ready and waiting for our episodes every Tuesday morning.

THE AGONY OF WAITING FOR LISTENERS

Let's say you spend weeks planning your podcast. You do all of the legwork, including creating a website, setting aside a recording area in your house, and finally recording and editing your first episode. You put it online and... crickets. There may be dozen or so people who download the podcast, but none of them have left any comments or reached out to give you feedback.

Now what?

There's nothing more demotivating for a content creator than realizing that a big audience just isn't there and interest in all of your hard work is low.

The good news is that, unless you are a huge celebrity who already has a large online following, this is totally normal. There is no way that you are going to have a significant audience right out of the gate. This will likely change and your audience will grow, maybe even faster than you think it will.

We found that our audience grew at a steady, consistent pace and although growth felt slow at first, it did increase at an exponential pace over the months and years after we got started.

Right after releasing your first few episodes is when you are going to have to practice the most patience. For our first few podcast episodes, we only had a dozen or so listeners, then two dozen, then maybe just fifty listeners - after we'd already spent hours and hours on our show, releasing at least 5 or 6 episodes by then. It hardly felt worth all of the effort we had put into the podcast for just a handful of listeners, and we could have easily gotten frustrated and given up in our first few months.

To get through this tough, early time when it felt like nobody was listening, I kept reminding myself to move forward. There were some things about podcasting that were motivating to me other than listeners. I was really enjoying the process of podcasting and everything I was learning throughout the process. I enjoyed the discussions Brittany and I were having, and I was rapidly improving my editing skills. Seeing my progress with regards to the content creation process was motivating. And importantly, we both kept each other motivated and excited to move forward.

Focusing on the things I was enjoying about podcasting made it a little easier to ignore the negatives, like low listener numbers.

But then, as the months ticked by, before I knew it, our first dozen, then two dozen podcast episodes had been released, and suddenly we had hundreds of listeners. We received a handful of positive reviews on iTunes. And at that point, I allowed myself to get excited about our listeners and started to focus more on their input and the positive feedback we were getting from our work.

Looking back on it all now, I believe we did build a considerable audience in only about a year. It was just that period, right when we were starting out and trying to stay motivated to keep going with what sometimes felt like a daunting project and only attracting a few listeners, felt like it lasted forever. So, more than anything, focusing on the positives and what I was enjoying about the process kept me motivated.

In short, simply stick with it and get through those more daunting early times - when you have fewer listeners to keep you motivated - by focusing on what you love about podcasting.

I should also add that there is no ideal audience size for a podcast. Celebrity-hosted podcasts may quickly garner hundreds of thousands of downloads thanks to their existing networks of followers, or the central host's popularity or recognition. On the other hand, niche podcasts about specific topics like, say, crocheting or hiking, may only ever gain a hundred or so regular listeners. Needless to say, avoid comparing yourself to others. Your podcast is not their podcast. I really do believe that 100 loyal and engaged listeners are more exciting, in my experience, to have involved with your podcast than a few thousand silent and passive listeners who never interact with you and keep their distance. That is why not putting so much stock on numbers of listeners, but rather developing a solid community of listeners, should be an important part of podcasting.

SEASONS

Since our podcasting schedule was in many ways pre-set for us by the television season, we did not have to make a lot of our own choices with regards to how to set up our "podcast season." However, many podcasters will need to decide when and if to take a hiatus from podcasting at some point, perhaps dividing up their show into "seasons": i.e., podcasting every week for four months, then taking a three month break, and then continuing on for another four months, and so on. Podcasting seasons can be very beneficial to you, as the creator, because the time in between will give you time to rest, reflect, and recharge. As with school semesters, which are broken into 4-5 month stretches punctuated by longer winter or summer vacations, podcasts can also have a specific season and then a longer break to allow you some down time away from the podcast.

If I had to create a new podcast, I would try to follow a similar schedule to the one we had with Once Upon a Podcast: we released new episodes once a week for a 3-4 month stretch at a time, followed by a short break of 1-2 months. In many ways, I believe it's a good idea to do enough episodes in a row to get listeners drawn in and hooked on listening to you on a weekly basis. So, the 3-4 month stretch of new podcast episodes on a weekly basis was enough to get listeners hooked and used to a routine of listening to us.

On the other hand, I also think it's a good strategy to not podcast so much that you burn out and your listeners grow tired of you. That's a worst case scenario. It's much better to leave them wanting more. That is why I felt podcasting in seasons worked quite well for us. If you have the freedom to come up with your own podcasting schedule - if your show is not tied up with any particular event or schedule set by outside entities- you should consider not only how frequently you want to podcast (i.e., once a week, twice a week, twice a month, etc.), but whether you would also like to schedule in any hiatuses or breaks from the podcasting season (i.e., taking the months of December, January, July and August off, for example). If you do decide to have a break, consider having a similar schedule each year so that in the long term, listeners can learn your schedule and anticipate it. If you decide in advance when you will be taking breaks, you can warn your listeners a few weeks in advance when your season is winding down so they aren't taken by surprise by your sudden disappearance. This will also give you time to clearly communicate when you will return so that they can anticipate that date well in advance.

There is no exact science to scheduling your seasons. You will have to take a leap of faith and choose a schedule (and decide on whether or not to do show seasons) that you believe will work best for you, and do not be afraid to change things up or try something new if that schedule does not end up working out for you (or your listeners).

If you are not sure which podcasting schedule would be best for you, I decided to offer a few suggestions, along with what I believe are the major pros and cons of each option. The options you could choose when it comes to a podcast release schedule are pretty limitless, but this might help point you in the right direction.

EXAMPLE SCHEDULE OPTION 1: RELEASE A NEW PODCAST EPISODE WEEKLY OR BI-WEEKLY, WITH NO SEASONAL BREAKS

If you decide to podcast on a regular, set schedule, from now until, well, forever, then it's hard for me to find too many cons about that from the point of view of your listeners. Once you gain listeners, the benefit is that they will always know exactly when to expect a new podcast episode from you and that predictability will make it easier for them to remain loyal listeners. Your podcast will more easily become a part of their weekly habits and they will clearly understand when and where to find your new podcast episode.

From the podcast creator's perspective, there are some plusses to this approach. For one thing, you will build the podcast into your routine and it will become a predictable part of your life. You'll get into the habit of putting out podcast episodes at the same time every week or every other week and you will never fall out of the habit of doing so after, say, taking a summer off from podcasting.

I will, however, caution that this is a huge risk from a podcast creator's perspective as far as your personal well-being goes. If you are opting for this schedule and podcasting is a part of your 9 to 5 job description, then this may be the way to go since it is a part of your career and you can, say, work it into your job's to-do list every week. However, if you are podcasting for fun or for a hobby, this could prove to be very draining and unsustainable over the long term, unless you have significant amounts of free time every week (or every other week) to devote to podcasting.

Brittany and I decided to try podcasting with this schedule when we launched our second podcast, "Pop Culture Detectives." As I've mentioned a couple of times now, this show ended up being a very short-lived podcast. We made it through only 20 or so episodes before we had to throw in the towel. We felt bad about abandoning it, but we were really left with no choice because doing those 20 podcast episodes, a new one every other week, quite simply burnt us out. We were doing this podcast on top of our Once Upon a Podcast, which was far too much, but even without our main podcast, I still think that Pop Culture Detectives ("PCD") would have burnt us out over time.

We went into PCD thinking it would be manageable to podcast every other week. But in reality, podcasting with no break in sight ended up being draining. It was mentally too overwhelming to commit to podcasting twice a month for "forever." As we got a few more months into PCD, we were slowly feeling drained, exhausted, and totally unenthusiastic about PCD. We found that we were spending that "off" week usually thinking about, researching, and promoting the PCD episodes anyways, so we never truly had any down time to refresh and recharge. Ultimately, we realized that in many ways podcasting every week, but with seasonal breaks after 10-12 weeks of podcasting, was a much better approach, which was what we had been doing for Once Upon a Podcast. The 10-12 weeks when we had to podcast every week sometimes felt like a marathon, but we rested and recharged during the months off that we had with Once Upon a Podcast.

By not having any downtime ever with PCD, it felt like more of a job than Once Upon a Podcast ever had, and we felt frustrated that it was never going to end. In short, this schedule was not for us, and we quit the podcast because we burned out. It would have been better if we'd picked a more sustainable long-term schedule in the first place, or at the very least, tried to modify it before we reached the point where we just didn't even want to think about it anymore and preferred quitting to trying to improve the situation. That's human nature: when the times are rough, we just want to run. This is the danger of attempting a podcast with no break: as I discovered, it will feel like another job, and you will have no time to rest, reflect, and regenerate.

In retrospect, I think that we could have given PCD another go, but with the caveat that we would have had to plan in seasonal breaks for our own mental sanity. We never ended up picking up the podcast again because we ran out of momentum.

Whatever schedule you choose, just be sure it is a realistic option for you and your life, and also be ready to adjust your release schedule if you do feel like you are about to burn out. Learn from our mistake and at least to make improvements before completely burning out.

Ultimately this was the schedule that I preferred for podcasting. Weekly podcasts, I found, helped ensure that our podcast did well as far as search engine optimization and podcast app algorithms go (as far as we know about the elusive search engine algorithms, search engines and platforms like iTunes prioritize podcasts that post new episodes regularly and frequently, and weekly seemed to fit the bill, as Once Upon a Podcast was routinely one of the top search results when searching for Once Upon a Time-themed podcasts on apps like iTunes).

Releasing a new episode on a weekly basis also enabled our listeners to get into that all-important habit of looking for, and listening to, our new episodes frequently. We had listeners comment to us that they were "addicted" to listening to us weekly, or that they looked forward to their Tuesday commute to work because of our show. This told me that podcasting weekly was frequent enough to become a part of the lives of our listeners, and helped us create a loyal listener base.

On the other hand, as I touched on above, having seasonal breaks was also a good thing for us. It allowed us some down time and rest away from having to podcast every week. We always found that by taking a summer or winter break from podcasting, we would come back into our new podcast season refreshed and with a renewed sense of creativity, often coming up with creative new ideas that we might not have thought if we hadn't had the down time. Taking some breaks enabled us to have the time we needed to actually be productive and creative when we were back at it during the season.

In addition, I always suspected that it wasn't necessarily a bad thing for our listeners to miss us. Many of our listeners would write to us during the break and mention things about how their commutes were not the same without us, or that they missed listening to us while they were cleaning the house over the summer break. This was not only nice to hear and I appreciated that our listeners missed us, but it also demonstrated that if you build a loyal enough audience, they will anticipate your show's return after a hiatus. Once we built a loyal listener base, they would keep track of the date when we returned and would be right there to download our new episodes.

(Admittedly, as a podcast about a television show, it of course helped that our podcast returned at the same time the television series did for their new season, which made it easier for our audience to remember to listen to us as well).

Regardless, I always felt that leaving our listeners wanting more of us was better than having all of our listeners simply get sick of us and move on to other podcasts.

EXAMPLE SCHEDULE OPTION 3: LESS FREQUENT UPLOADING SCHEDULE, SUCH AS MONTHLY, OR NO SET SCHEDULE

This option is where it's all about balance. Ultimately, as a podcaster, you will need to balance the frequency of your new episode uploads with your own personal sanity and ability to put out quality, creative content over the long run. Although having a less rigid or frequent schedule will give you more space to breathe, edit, do more creative work with your podcast, and possibly come up with more new and creative ideas than podcasters who get stuck in the rut of putting out a new episode every week, there are also some dangers when it comes to having an infrequent upload schedule or no set schedule at all.

Some of the best, most high-level podcasts I have listened to are ones that come out only every month or less. Their creators work hard behind the scenes to produce high quality, thoughtful and original content, and quality does take time. A less frequent upload schedule might enable a creator to put much more work into their podcast. In these cases, listeners may find it well worth a long wait between podcast episodes. Podcast episodes that are released less frequently can be more heavily edited, can incorporate unique elements like original music, can feature special guests, and so forth. If you have a lot of very creative ideas for your podcast that may take time to build into each new episode, then giving yourself lots of time and space to create an episode may be worth it. You might also gain a loyal listener base if you do a very good job with your show, as they may find it worth the wait for each of your new episodes.

Podcasts that are done in a more narrative or documentary style often have a longer wait between each episode. They might have many different segments, interviews, and background information and research that goes into each episode. Obviously, something like that cannot be turned around quickly, as content needs to be gathered over time and editing may take a while.

On the other hand, breaks that are too long, and too frequently taken, can pose challenges in terms of both you having a podcast routine and your audience remembering to listen to you. From a creator's perspective, not having a regular schedule may actually cause you to become too relaxed and slack off from creating new episodes, and before you know it, 6 weeks will have gone by with no finished episode.

It may be difficult to build a loyal base of listeners if you are not podcasting very frequently. From your listeners' perspective, it might just be too confusing to understand when you release new episodes, and they might wait and wait and move on eventually because you haven't released anything in a while. Listeners might also forget about your podcast. They may also never have a chance to get into the habit of listening to you because you don't upload frequently enough. This can all create barriers in terms of gaining a loyal listener base. To try to prevent some of these issues, you may still want to schedule your podcast episodes - such as setting a release date of a new episode every 1st of the month - so listeners could still have an idea of when to look for your new podcast episodes.

My experience taught me that I preferred podcasting frequently enough to build a loyal listener base and to get into the habit and rhythm of consistent, frequent uploads, but to schedule in long enough breaks to allow myself time to rest, reflect, recharge and come up with new ideas. Your results and preferences may vary. The most important thing is to be flexible and also be aware enough of how your schedule is working for you, and to be willing to change if you are having problems.

Like anything else in life worth doing, podcasting is hard work, but rewarding. Look for, and celebrate, small successes. Evaluate, and re-evaluate, what you love most about podcasting, and try to cultivate that as much as possible, so that you don't give up prematurely. Building an audience takes time and it takes a while for them to learn when and where to find you. Most importantly, choose a schedule that works for you, so that you can happily and successfully podcast for a long time.

CHAPTER 9: DEVELOPING A COMMUNITY

One of the most unexpectedly rewarding, and also tremendously necessary, aspects of building a successful podcast.

I am writing this chapter as I just got back from an outing with a friend... a friend I made because of Once Upon a Podcast.

One of the most surprising things I found with podcasting was how naturally a community formed around our show. As podcasters, we both worked to foster a community, but in the end, it also just sort of happened. A group of loyal listeners kept returning to our podcast each week and before we knew it, we knew them by names, we knew a bit about their lives and families, and we even met some of them.

s without saying that by the year we began our podcast, in 2013, many people had smartphones, and along with their smartphones, social media accounts. What had been just a few years prior to our podcast a world filled with geeks and nerds, social media was on the rise in the early 2010s and used by more and more "ordinary" (meaning, non-nerdy) people around the world. Promoting our podcast on social media was a necessity when we started the podcast because, by then, most people were on social media and used it daily (or rather, hourly). By the time we ended our podcast in 2018, thanks to our presence on social media, we had built up a community of followers on Twitter, Facebook and Instagram. (We also had Snapchat followers for a hot minute, until everyone migrated to Instagram Stories).

Of course, when it comes to social media, one thing might be the biggest thing one minute and out the next (hello, Snapchat). Any content creator, podcasters included, should be flexible, open-minded and always ahead of the "next big thing" to make sure that you can reach and interact with your audiences where they are. Technology moves quickly, and so it's better to not overthink things when it comes to social media: just sign up and start using it. It might become the next indispensable thing like Facebook, or it might be passé in a week, but the key is to stay ahead of your audiences.

When we started our podcast in 2013, most other podcasts that we researched published new episodes of their podcasts as a part of long, narrative, photo-filled blogs. (I could be wrong about this, but I always think of the years 2010-2012 to be the "glory years of the blog", when everyone and their mother had a blog. But that is a totally unscientific claim... it's just my own memories of those times!)

We did start a blog and even posted some narrative written content to it, but by the end of our podcast run, we didn't have any visitors to it and we had stopped using it for anything other than to post each new podcast episode itself (which was fed to podcast platforms from our blog). However, in 2013, many other podcasts at the time were not only posting their episodes to a blog like we did, but were using their blog as a destination for fans of their podcast to read news and other fun posts related to their interests. Listeners in 2013 were, based on my observations at the time, still seeking an actual website to visit that had information and extra content to consume, such as news and pictures. In more recent years, social media has filled in that role and large websites and long-form blog posts are less sought after.

There was another benefit to having a website and a blog, however, that should be noted: having a website containing information and blog posts related to your podcast boosted the name of your podcast and your podcast's website in search engine rankings. Google used to favor websites that posted new content frequently, giving podcast creators an incentive to maintain a website for their show and update it regularly. In fact, early on in our days of podcasting, we felt that we might only be able to drive listeners to our podcast if we made our website really content-rich and posted news and other topical items about the television show that our podcast was about. Regularly posting a lot of new content relevant to the topic of your podcast is one way to boost your website (and by extension, podcast), in search engine results like Google. This is called Search Engine Optimization. While Search Engine Optimization (SEO) is still important for podcasters to understand and learn about, today I believe it is less common for listeners to find podcasts via Google than it was when we first started podcasting in 2013.

Back then, so few "average" people knew about podcasts that podcasters had to draw in potential listeners to their website and then lure them towards their podcast. Now, so many people use apps and smart home devices such as iTunes, Google Play, Spotify, Alexa, etc. to find a podcast directly, and would rather consume content via podcasting platforms or social media platforms than visit individual websites first for new content, that it is less likely that a significant number of potential podcast listeners will be found by first enticing them to visit your website. Even a couple of years into our podcast, we no longer had very many listeners visit our website or find us on search engines like Google. By 2017, our listeners generally found our show directly through podcasting platforms like iTunes.

It is worth reading up a bit on SEO before starting your podcast. SEO can help you understand more about how potential listeners could find your podcast and how you can optimize things like your website or social media profiles to help potential listeners discover your podcast via search engines. SEO is worth understanding but it also changes frequently, which is why I am not going into it in great detail here and instead recommend looking for some current (within the last 6 months) articles or guides on the topic. Search engine algorithms can change so quickly (and indeed, no one knows exactly what algorithms Google uses to rank their pages anyways,) that it is best to do your research right when you are first starting your podcast, and then staying up to date on trends whenever you have some time to spare.

From websites and blogs in 2013 to podcast-specific apps in 2018 (and our varying experiments with Twitter, Snapchat, Instagram, Tumblr, and Facebook in between): the way our listeners found our show changed pretty significantly during the course of our podcast's life. As you read this book today, how the majority of podcast listeners might find your podcast may or may not be the same as it was for us in the late 2010s. Do some research, read up on how podcasters are reaching audiences right now, and stay up to date and aware of these trends and watch as they change. In all likelihood, the means in which audiences find your show today could change again as quickly as 6-12 months from now. If you're not sure how to stay up to date on all of this: simply observe. Follow other podcasters and influencers, see what they are doing and how they are advertising their content and connecting with their audiences, and explore whether that approach might be right for you and your show.

I believe the most important skill when it comes to publicizing your podcast is being flexible, open-minded and willing to quickly change. If we had been married to the idea of posting new blog posts five times a week for the five years of our podcast, we would have likely wasted our time, as fewer people read blogs in 2018 than they did in 2013 and that would have probably not attracted as many listeners as our various experiments with social media and making sure our podcast was available on popular apps did.

Instead, we shifted to publicizing our show to a potential audience via various social media platforms, and by the end of our podcast run, found that connecting with our listeners through communities and groups on Facebook was the most efficient way of promoting our show. Going from promoting our podcast and the topic of our show on a blog/website to Facebook Live videos and using Facebook groups to communicate with listeners was a fairly significant shift, but that is what happens in five years. Five years from now, the way podcasters can best reach out to and connect with listeners will have likely changed again... possibly even more than once.

COMMUNITY DEVELOPMENT 101

One thing was consistent in the five years we podcasted, and that was the importance of connecting with people and building a community. The means may change, but the end goal was the same: connection.

This is a fairly abstract subject to try to discuss and there really is no one hard and set rule for developing a community, but there are some characteristics you can foster within yourself and steps you can take to improve your chances of building a community. Being open-minded is key. Compassion and empathy are also essential "human" skills to have a good grasp of. Emotional intelligence will support you and your goals to connect and build meaningful relationships with others.

You also need to be willing to connect with individuals on the platforms they can actually be found on. I once heard a small business owner complain that they hated Facebook and would never start a Facebook account for themselves or their business. I think that is a pretty self-sabotaging choice, because even if you dislike Facebook, millions of other people use it almost every hour of every day, so if you want to run a successful podcast (or business), making sure you have at least some presence on there is critical. It actually stands out if you don't have a Facebook page or group for your show (or business).

Community building is, in my mind, fostering spaces where people of similar interests can gather and connect. We worked hard to increase the chances that like-minded people with similar interests could find each other by fostering the spaces where they could connect with us and each other (and, of course, connect with and discuss our podcast, too).

If you'll recall the earlier chapters of this book about how to begin with a strong concept and then find your voice, remember that I suggested you consider the type of audience you want to attract to your podcast. Since you have already thought about the types of people you want to attract to your podcast, you should now explore the virtual spaces where those people might be found. Knowing your audience should help you also understand which social media plat-forms where they are most likely to be found. Do a bit of re-search: if your podcast is geared towards younger audiences, consider signing up for accounts on the latest and trendiest social media platforms, learn to use it in the same way every-one else is using it, and cultivate a personality and profile of your own on there.

This can also be a process of trial and error. When we first started our podcast, we believed a lot of our listeners could be found on Tumblr, as there was a fairly significant fan base for Once Upon a Time on there. Although we managed to attract the attention of a few listeners on Tumblr, we quickly found that not a lot of traffic was driven to our podcast from Tumblr. So, we moved on to other social media platforms, especially Twitter and Facebook, which ended up being our biggest places to connect with audiences. Even Twitter changed, however. Although we found a lot of people who were fans of the show on Twitter in 2013-2015, the U.S. election season in 2015-2016 began to change the tone of Twitter. The social media platform became very politicized around the elections. As it became more political, people who had once gravitated to Twitter to discuss fun topics like television began looking elsewhere for their entertainment fix, as the social media site had become permeated with poli-tics. We found that over 2015-2016, we had fewer interac-tions on Twitter, and as a result, focused on our podcast's profile there a little less. Many of our listeners gravitated to Facebook, so we started to put more energy into our pres-ence on Facebook.

Despite this shift in terms of which platform was generally used more often by our listeners, we maintained our social media profiles on all of those sites: our Tumblr profile at the time helped boost our Google SEO, and our Twitter profile at the time remained somewhat of a useful tool for some of our listeners to send us messages. We never abandoned any of our social media profiles because all were useful in some way.

CULT OF PERSONALITY

Over the five years of podcasting, we also noticed a trend in terms of how social media was used. When we first started our podcast, most podcasters had social media handles for the podcast itself. The avatar for their profile might be their podcast logo, and the handle might be the name of their podcast (for instance, our Twitter handle was @onceuponpodcast). The names of the podcast hosts themselves seemed to remain somewhat more anonymous, and podcast hosts even used pseudonyms on those platforms, not connecting their real identities or personal profile pictures with their show.

However, by 2018, a big change had happened. We noticed that most of our listeners wanted to connect with us, the hosts, the "real" people behind the show. In the latter years of our podcast, we found that our listeners were following our personal accounts on Twitter and friending us on Facebook, rather than following our podcast show's "official" social media accounts. It felt like the concept of gathering everyone around the "brand" of your podcast had become outdated, as everyone instead wanted to feel like they were connecting with you, the host, the person they heard talking each week. Which makes sense, if you think about it: humans really just want to connect with other humans, not inanimate podcast logos or branded accounts.

By the end of our podcast run, Brittany and I had a lot of our listeners following our personal accounts and we began to become more active and engaged on our personal accounts rather than on our official podcast account. That

trend seemed to encourage a warmer and more personal approach to publicizing our work, and allowed us to connect
even more closely with our listeners.

Today, I believe this trend - of connecting with other humans, not with sterile branded accounts - remains on most
social media platforms. That means that you may not even
need to create a social media profile for your podcast itself:
instead, you could direct listeners to your own personal social media profiles, and vice versa, on your social media accounts, you should integrate news and updates about your
podcast.

I suggest making your own personal profile where you
think your ideal or target audience might be found, whether it's on Facebook, Twitter, Instagram, Pinterest, Tumblr,
WhatsApp, Reddit, an automobile repair website, a message
forum dedicated to a 90s tv series that itself has been around
since the 90s, a local knitting club, or just the bulletin board
in the coffee shop down the street. No matter what the space
is, virtual or "real," you should join, become a part of that
community, build your identity within it. Only when - and
if - you have built a presence there as a user and a member of
the community, can you start to think about ways in which
you can branch out and use your presence in those arenas to
promote your own work (respecting community guidelines,
of course). Promoting your podcast within the communities
you are a part of, according to the rules and norms of the
community, will come more naturally after you are familiar
with the space and your relationship within it first.

The social media platforms, websites or other areas where
you are working to build a community for your podcast or
participate in a community that may be interested in your
podcast may shift in nature or change in just a year or two.
That's why it's a good idea to dip your toes into multiple

places whenever you can. Make it a habit to check out new social media platforms or visit sites that seem to be driving traffic to your podcast (or places where you hear or see other podcasters promoting their own work). Don't overthink your presence on these sites or social media apps: just register, use them, become familiar with their features, and see if it's a space that might be good for building a community. Again, it's all about being flexible and curious and going with the flow, because online trends shift quickly.

Social media, a history

To give you a better sense of how rapidly social media and online community trends change, I thought it might be fun to give you a quick retrospective of our own evolution in promoting the podcast online, to provide some sense of the variety of options out there, and how much current trends could change over the years you will be podcasting.

When we first started our podcast in 2013, many other podcasts (on topics similar to ours) had their own individual websites, complete with blogs, information, photos, artwork and more. Some podcasts even hosted their own separate message board communities, where their listeners would sign up for an account and be able to post about their reactions to the podcast episode or other thoughts on related topics.

This was, no doubt, a lot of work for podcasters. Hosting and maintaining a full website and individual message board communities was a large task in of itself, let alone doing that on top of putting out a podcast. Taking on such a project would have been expensive, time consuming and probably require more than just the two of us. Although we did, in the early days of our podcast, build a website that had a blog and some other information on it, we felt that a very robust website with photos, artwork and message boards was too much for us to take on and was even starting to seem a little outdated at the time, so we didn't end up building a very complex website to go with our show.

The first social media account I created for our podcast was a Twitter account, and my co-host created a Facebook page for us. These two accounts ended up being our biggest promotional tools and helped us the most with building our community of listeners over the years. Both were also interesting adventures for us and both platforms, fortunately for us, continued to grow significantly in the five years of our podcast in terms of number of users and general popularity with the public. Although we found that fewer of our listeners connected to us in the later years on Twitter than had in the earlier years of our podcast, Twitter still remained a consistent source of connection with our listeners, as it was one of their favourite ways of reaching out and communicating with us.

In fact, when we first started our podcast, it just so happened to be a year when it was popular for television shows to promote themselves on Twitter. We noticed that many cast, crew and writers from tv shows, including Once Upon a Time, were actively using Twitter as a space to promote themselves and the show. Twitter still felt a bit like a "Wild West" as far as social media platforms went; even though it wasn't new, it was only just gaining mass popularity in that 2012-2013 era. As a result, many of the actors' tweets were impromptu and a little bit more casual, which resulted in a friendly, laid-back atmosphere. Once Upon a Time even seemed to encourage (perhaps even require?) their actors and cast members to be on Twitter and talk about a new episode of the show as it aired. This resulted in many of the cast members of the show "live tweeting" their thoughts and reactions to the episode in real-time while the show was airing on network tv. Fans took notice and logged onto Twitter every Sunday night while they watched the new episode to see the "behind the scenes commentary" that several of the cast members would provide on their Twitter feeds in real-time.

As a budding podcaster, I quickly saw the potential of this atmosphere. I started to also "live-tweet" while the new episode aired, posting my real-time reactions as they happened and, as hoped, my tweets started to get attention from the mass amount of fans that were on Twitter while watching the show. I was retweeted by other viewers of the show. I also had my tweets "liked" or "retweeted" from time to time by someone connected to the show itself. This attention that my Twitter account received every Sunday night while the show aired prompted Twitter users to click on our Twitter profile and follow its link to our podcast episodes. At some point, probably as a result of all of my live tweeting activity, Twitter gave our podcast's Twitter account enhanced analytics features, which better allowed me to track the success of our social media "performance" each week. I could see how many people viewed or interacted with each Tweet I had made, and which Tweets prompted people to click on our profile or an attached link, and so on. Several third party websites (most are no longer available today) also provided me with additional insights on how each Tweet performed and how many people viewed our Twitter profile or followed the link to our site each night.

I could not count on my live tweeting to drive listeners to our podcast forever, however. After a year or so, many of the cast and crew of the show stopped live tweeting the episodes as frequently. This actually gave me a bit of a boost: with no cast members doing the regular live tweeting, but fans still in the habit of opening up Twitter during the show to "watch" it with other fans, the usefulness of my live tweeting persisted for a while longer. More and more fans started to live tweet the show themselves, using Twitter as a giant chatroom of sorts to discuss each episode, and many of the other podcasts about the show also followed suit and began to tweet during the show, too. In fact, this was when I really started to connect with some of our listeners, especially those who were on Twitter during the show every week and were following me while I was live tweeting. They would reply to my Tweets. The back and forth dialogue I had with many of our listeners during the episode resulted in connections and conversations that had a lasting impact. I began to build relationships with our listeners during these times, and likewise, our listeners started to find each other on Twitter, too. Suddenly, a community was forming.

As with everything, however, trends shifted and the fandom's live tweeting enthusiasm started to wane. I saw less and less interest in the live tweeting during the show about three years in, and by our last year of podcasting, it hardly garnered any interest. This "golden age of live tweeting" had come and gone in the space of about two and a half years, but it left a lasting impact in terms of our listener community. Some of our Twitter followers migrated to other social media platforms that were becoming more popular for entertainment content, such as Facebook and Instagram.

I also had an interesting experience promoting our show via Facebook. While I was largely in charge of our Twitter content, my co-host did a lot of the legwork with Facebook. (Again, this is the definite plus to having at least one partner when going into a podcast: you can split the tasks.)

When we first started our podcast page on Facebook in 2013, it took ages to get to just thirty "likes." However, all of the sudden, somewhere in 2015, Facebook exploded in popularity. It seemed like everyone and their grandma (literally) by that point had a Facebook account, and a lot of communities had really began to grow on Facebook. Many of the internet's old fashioned message board-style communities had largely migrated to Facebook by then, and new people were joining Facebook groups related to their interests. Finally, our Facebook group was getting more and more likes, and with this, we were in turn inspired to use Facebook more heavily as a place to discuss our podcast and the television series itself.

My co-host began to post more discussion prompts and even posted tidbits of news about the television series on our Facebook page, to keep conversation flowing throughout the week in between podcast episodes. The reaction to this was good: we found that listeners were replying to our prompts and even sometimes going to our Facebook group as their source of news about the television show. Many of our listeners took over the conversation, and began posting discussion starters and news items of their own to our Facebook page, which we of course welcomed as it helped build the community and keep the group active. This was much more efficient than having a standalone website be a source of news and additional information: Facebook enabled us to essentially crowdsource this task of posting news, and because "everyone and their grandmother" has a Facebook profile, it was a better environment for conversation to flow, not to mention a place that a lot of people logged into every day to check. In other words, social media usage had evolved right before our eyes (Facebook groups in 2017 were so much bigger than they had been in 2013). Facebook became, in the end, a critical part of how we built and fostered a loyal and dedicated community of listeners.

We experimented with a few other social media platforms,

including Instagram. This is another one of those cases in which the use of Instagram evolved in a way that wasn't necessarily what we had planned it would, but we rolled with. We started on Instagram with a profile for the podcast itself. However, with the shift in listeners wanting to know about us, the real people behind the show, we found that many of our listeners ultimately followed our personal Instagram accounts rather than our podcast account.

We also looked for new opportunities all of the time. When Facebook Live first launched, my co-host and I immediately tried it out on our podcast group page. Our listeners loved the Facebook Live video, and so we started to do one every week. In the first few weeks of Facebook Live, Facebook had not perfected their search algorithms - to our advantage. We were actually appearing in the search results for Once Upon a Time on Facebook, and many of our videos were consistently ranked as highly as videos from the official show's Facebook page. This drove a lot of traffic to our Facebook Live videos. With Facebook promoting our videos at the same priority level as the official show's videos, this gave our podcast a big boost in the Facebook search results and drove a lot of people to our podcast's Facebook page, which by extension drove new listeners to our podcast. The algorithm changed after a few weeks and we were no longer as favored in the rankings, but we made the most of it while we could. At least we got lucky with Facbeook's search algorithm for a while.

Still, doing live videos on Facebook was very popular with our listeners for a few months and I enjoyed doing them. Once again, it was a away for me to get to know many of our listeners and have real-time conversations with them. It also helped them to get to know us better, especially since they had only known our voices previously, and Facebook Live was a video of myself, so it was another aspect that helped them get to know us a bit better.

Although we strove to go with the flow when it came to promoting our podcast, connecting with communities, and seeking out trends, on the other hand, we didn't always blindly follow every trend. We were strategic about how we used our time and energy. We didn't jump on every bandwagon (for instance, we never saw the need to have a Pinterest account, nor did we ever really fall in love with Snapchat despite our listeners pressuring us to join). We tended to do things we loved: live tweeting on Twitter, doing Facebook Live on Facebook, posting news and conversation prompts in our Facebook group. In the end, we had to be true to ourselves in terms of how we connected with our audience and promoted our show online, because that is how we never grew tired of doing it, and also, I believe that is why our listeners started to feel a more genuine connection with us. We were doing what we clearly loved.

THE REWARDS OF A COMMUNITY

At the end of the day, the online platforms where potential listeners - and people in general - will congregate is going to change and evolve. It may evolve a month from now, or eight years from now, but nothing tends to stay too consistent for too long.

What does stay consistent over time is that people are drawn to online communities of other individuals with similar interests. As outlandish as it might sound to someone who is just starting a podcast, there may be a day when a community forms around your podcast. We found that a core group of individuals started to comment frequently on our podcast Facebook page. Eventually, they started to talk amongst themselves in comments and replies to our discussion-prompt posts. After about a year, they had become friendly with each other (at least through Facebook) and to this day I am still Facebook friends with several of them. In other words, our little community actually outlived the podcast itself!

A community doesn't have to consist of every one of your listeners, or even hundreds of people. What it can consist of is a core group of individuals who genuinely enjoy your podcast. They may spread the word about your podcast to others, recommend the podcast on other social media platforms or in other communities they belong to. They may also write positive reviews of your podcast on directories like iTunes, or repost your social media posts. Their contribution from a marketing point of view is invaluable: their genuine interest and enthusiasm about your podcast might attract the attention of others and be the reason someone else gives your podcast a try. But they don't have to do any of that, either. Just having a listener who was engaged with our podcast was very rewarding in of itself.

Being a content creator is a lonely process and it is often a long road to get to a point where you have attracted just a small handful of regular listeners. I found it immensely encouraging every time I connected with a listener. When I started to question what I was doing and why, the genuine engagement and positivity that others showed towards our podcast was probably one of the most significant motivators. It helped me realize that what I was doing actually had an impact - however small - on other people, helping them get through a long commute or work week, or breaking up their boredom or loneliness.

I also learned so much from many of our listeners. Sometimes their feedback or reaction to a comment I had made on the show would be deeply thought-provoking to me. Thanks to interactions and comments we received from listeners, I also learned about different areas of the world, new topics I hadn't thought of to explore, and gained new perspectives on the material we discussed each week. That is the benefit of putting your work online: you will reach a global audience and, as such, are almost guaranteed to learn something from that alone. Being a podcast creator is a wonderful way to learn and grow as a human being.

HOW DO I KNOW HOW MANY PEOPLE LISTEN TO MY PODCAST?

One of the most frequently asked questions I see from podcasters is… how do you even know how many people are listening to your podcast?

Unfortunately, the only answer I have is not one you will find easy to hear: there really is no good way to know for sure.

A few years ago, Apple said they would start providing podcast statistics in iTunes. Despite that being a few years ago, there still is no such tool available to podcasters who list their show on iTunes. And even if there was, keep in mind that the statistics would probably only cover the number of downloads that the podcast receives by people using Apple's apps to listen to podcasts, which has limited usefulness, as many podcast listeners seek podcasts on Android devices, on Spotify, on smart home devices like Amazon's Alexa or Google Play.

The only way I had a decent idea of how many people were listening to our podcast episodes week to week was by "guesstimating" according to the tools that were available to me at the time. Likewise, you will need to do some research to see if there are any tools available to you based on the way in which you get your podcast online in the first place (for instance, whether you self-host and upload your podcast onto a website and use a tool like Blubrry to feed it to pod-cast directories, the way I did, or if you use a podcast hosting service that may offer their own statistics tools).

The way I calculated how many listeners we had was via the stats tool that Blubrry had on their website. I had a Bluberry account and could log into my account and take advantage of a few of their free tools, which included a basic statistics fea-ture. However, even this was an imperfect method of find-ing out the number of listeners we had, as Bluberry's website itself states that their statistics are approximate.

At the time we had our podcasts, Blubrry measured "down-loads" of a podcast episode, and not necessarily the amount of times an episode was listened to. For instance, using downloads as a statistic doesn't always give the full picture, since some users may set their podcast app to automatically download a range of shows that they then do not even listen to. There are settings on Apple's Podcasts app that cause the app to automatically download every new episode of every show a user has ever subscribed to automatically, but that does not mean that person actually listened to all of those episodes. Or, some listeners may actually download the same podcast episode many times; for instance, if they download the episode to their phone to listen to during a commute, but then get to work before finishing the episode and decide listen to it on their work laptop's browser. They are actually "downloading" that podcast episode twice, even though they are only one individual listener.

To further complicate things, some podcast platforms host audio and share it from their servers, rather than the server that you use to store your podcast episodes, making it harder for you to track how many people have listened to that same episode because it was through someone else's server.

Stats like Blubrry also count any download as long as it

took place, even if the episode was just partially downloaded (for example, if a person starts listening to your show for 15 seconds, but then stops the download and doesn't listen further, their 15 seconds of listening will still register as one download of that episode, the same as someone who truly listened to the entire episode). In addition, if a listener goes to your website and downloads the podcast directly through there and not through a podcasting platform where the episode was fed to, that download may also not register on any statistics tools.

In short, there are many ways in which stats are not scientifically accurate. When we were podcasting, I found that I was able to get a decent, though far from perfect, estimate of how we were doing and how many listeners we had for each episode through the following combination of tricks:

> • Early on when podcasting, we set up our RSS feed via Feedburner. Feedburner gave us some basic statistics for free. Feedburner was ultimately of limited usefulness to us, though it was interesting to see how the statistics we got from them varied dramatically compared to statistics we got from other sources.

> • Blubrry was the Wordpress extension that we used to get our podcast up and online, as well as fed to podcast platforms. Their free statistics tools provided some basic stats to us on their website, and had some interesting information, such as number of listeners per episode over different periods of time and the country or location of listeners. They also had additional statistics available for a small monthly fee (which I never

tried).

• If you choose to have your podcast hosted by a third party podcast hosting service (instead of self-hosting your show), some of those providers will offer their own statistics tools.

• If your podcast is self-hosted, some website hosts (such as ours, Dreamhost), offer statistics on how many GBs of material have been downloaded from your website server each month. By looking this up and doing some rough math (dividing the average file size of each podcast episode uploaded that month by the total GB of material that was downloaded from our entire website that month,) I could get an extremely rough, unscientific understanding of how many times our episodes had been downloaded that month. This was useful to essentially compare how much traffic we had month to month.

• Check out your show's position on podcast platforms like iTunes to see how your show compares with others in your podcast's category, or how you appear in the results (high or low) when you search for a key word related to the topic of your podcast and is likely a keyword used by other very similar podcasts as well. The higher you rank in your category or when you enter a popular search term, you can assume (typically) the higher your listenership is compared to other podcasts that are similar to your own.

Unlike media that is delivered via a single platform, such as a YouTube video or an Instagram photo, it can be difficult to understand how much attention your podcast and its individual episodes are earning because of the fact that your podcast will most likely be available via many different outlets and platforms. At the end of the day, though, not having a solid set of statistics to look up every week was a good lesson for us. It taught us that the number of listeners was not the most rewarding or important part of being a podcaster. Instead, we judged our success on the interactions we had with our listeners, the comments and feedback we received, and overall the rewarding experience we had building a community with them.

CHAPTER 10: AUDIO & VISUAL

Music and artwork are fundamental aspects of creating a podcast. Having a podcast involves building its brand. Artwork and logos will help your audience immediately recognize your show when scrolling through a podcast platform or app. Likewise, any music you use in your podcast, such as at the beginning or conclusion of your episodes, will start to become fundamentally associated with the overall feel and uniqueness of your show.

By this point, my hope for you is that you have built a solid foundation for your podcast and have a better sense of everything - both tangible and intangible - that needs to go into creating a successful show. Before I send you, and your carefully crafted podcast, off into the sunset, however, there is a bit more housekeeping to do. As a part of your role as a content creator, you will also need to create cover art, avatars and logos for your podcast, as well as think about whether you will use any musical elements in your show.

For those who are familiar with general digital content

creation - websites, videos, social media profiles, and so forth - the elements covered in this chapter may be no surprise. If you are new to putting your work online in any formal way, this chapter will help you get oriented with these two other crucial components of your show.

MUSIC

First up, and possibly the item that is most relevant to podcasters in particular, is music. As a podcaster, you will likely want intro music to start off your podcast episode, bumper music (the music that is played as an interlude to transition from one segment of your show to another), or outro music (to end each podcast episode). Music can become associated with your podcast. As an audio medium, any music you use will likely become a part of your show's brand and add to its overall feel and tone.

Before I continue onwards with my discussion about music, first a quick disclaimer. I am not a legal professional and am not qualified to tell you how to use music legally and responsibly on your podcast. I am not a legal specialist nor am I pretending to provide any legal advice at all, of any kind in this book. What follows in this section is purely my own experience with finding podcast music and using it on my podcast. Rules, regulations, and the availability of music that content creators can use in their work changes all of the time, and often also depends on where you live, where your audience lives, and so forth. So again, this is my own personal experience and my own personal context on the subject, and what may work for you might vary according to rapidly changing trends, current and local rules, regulations, laws, and other types of requirements. I strongly suggest you seek professional legal counsel if you have any doubts about using music on your podcast.

My first experience with copyright law was when I was at my first job, which was an entry level position in marketing. My boss wanted me to write to the organization that held the copyright on a song she hoped to use in an (internal) corporate video that would be handed out to employees only. We needed to get the rights to use that song in that video, pay whatever fee it entailed, and then distribute the video to our colleagues. Sounds simple, right? What followed were months and months on end of me basically failing at that task. At one point, it felt like a bad joke. No matter what I tried, how many people I talked to at that organization, how many forms I filled in, I couldn't get us the rights to use that (very old and obscure) song in our internal corporate video. Doing so would have cost thousands of dollars, involved a lot more paperwork, and really just required a lawyer or some other type of professional, which we did not have the budget or time to do. Needless to say, we dropped the idea of using that song in our video. I don't think we even ended up using any background music in that video at all.

Long story short, you cannot just use any song that you might have heard on the radio in your podcast. (Or on an internal corporate video.) Getting the rights to do so is also probably not a realistic option for most podcasters, as going through that process is time consuming and costly and possibly requires a certain set of expertise (or the engagement of an expert). You cannot use music by artists like Taylor Swift or Michael Jackson in your podcast. These artists' works are owned by other people, and to use their work you'd probably have to have a massive budget. Not to mention, the time and resources to seek and obtain permission to use the song in the first place.

The second thing to keep in mind is that you can't use just a little clip of copyrighted music. Every second of a song is copyrighted. So a 5-second clip cannot be used, just like a 30 second or 1 minute clip or the whole song. Many people believe that you can do this under fair use laws, or laws that allow for some copyrighted material to be used for educational or commentary purposes, or it is possible to use music if the podcast is not for profit (as long as you are not advertising or making money off of the podcast yourself). In my experience and research, none of these scenarios allow you to use music free of securing permission. My findings have been that, unless you hire a lawyer to help you understand and appropriately use of fair use laws, it is much better to just avoid using any music that is copyrighted altogether. I found it was a good idea to go by the rule that there is no cut and dry, legal way to use copyrighted music in your podcast, even in small (brief) amounts, or for educational purposes, and so forth. Because I wanted to sleep at night, we were very careful about not using music.

I also notice that a lot of content creators online believe that they can use material like songs, as well as reposting things like photographs or artwork on their website, if they give credit to the original creator or owner of the work (i.e. provide a link to the creator or owner's website). This is such widespread practice that many people have come to believe that using copyrighted material is fine as long as you attribute it. In most places like the United States, however, it is not legally permissible. The law tends to side with the original creator and to be safe, you should secure the permission of a creator, artist, author, etc. before using their work win any way.

On the bright side, there are many legal and totally realistic ways to use some kinds of music in your podcast. While you probably won't have any legal way to use a Top 40 hit in your podcast without going through lots of time, effort and expense to secure permission to do so, there are quite a few music libraries available to podcasters and content creators who need music for podcasts or videos. Over the years, this "safe" music for podcasters has been dubbed "Podsafe" music, and today is usually just referred to as "royalty free" music, as it is used by many other types of content creators as well.

Royalty free music for content creators

Royalty free music is music that is owned by the creator of the work, or someone else, and that owner licenses the music to an outside party to use that work in any way that they wish in their own work. Usually within the context of music for content creators, royalty free means that each time someone listens to your podcast, you do not have to pay the owner of the song a certain amount of money for that listener to hear the audio. Royalty-free material, even though it might still be copyrighted or protected by other types of intellectual property rights by someone else, may be used without the need to pay royalties or license fees for each use to that owner.

Copyright free music can also sometimes be found and used in podcasts. Copyright free means that the copyright on the song itself either expired or another party acquired the right to that copyright (and therefore the right to use that music in any way they want to).

In the years that I podcasted, I found that it could be frustrating (even fruitless) to try to track down free, Podsafe music and also verify that the Podsafe music that I found and wanted to use really was totally free, and not just stolen and posted on a website somewhere. I have found that rather than wasting a lot of time trying to find "free" music, paying a small fee to use music that is clearly indicated by its original owner and creator as royalty free was a much better approach.

During our Once Upon a Podcast years, I often visited a now-defunct royalty free music site for some music tracks that I usually paid a couple of dollars for (usually less than $5) in exchange for using that music as much as I wanted to in the podcast. The good thing about that site was that the music was straightforward as far as the usage permissions were concerned, and the music was affordable, with each track only costing a few dollars. I found it was a good resource for bumper music and outro music. The bad part about that site was that even in 2014, when we first started podcasting, much of the music sounded dated, as though it was from the 1990s or early 2000s (my cohost sometimes referred to it as "weather report music," the kind of music you might hear in the background during weather forecast reports). It was disappointing to not have nicer, more current-sounding music, since music does really reflect upon a podcast and set its tone and feel. Also, that site was, at the time, a well-known resource for podcasters, who had very limited choices available to them when it came to music, so many of the best tracks they offered had already been used by other podcasts.

Presently, royalty free music resources are on the rise, thanks to content creators who are constantly demanding more unique, up-to-date, fun music options to use in their videos, social media posts and, yes, podcasts. As I am writing this, there are so many more resources for music that I did not have when we started podcasting. Newer royalty free music resources have massive databases of songs. Typically these services are subscription based, though many sites also offer possibilities to pay a one-time fee to purchase a music track and use it royalty free.

The good thing about all of these service providers is that they were usually created specifically for content creators like podcasters, so they understand the needs that creators have and generally their music and services are tailored to those needs. Several popular services provide music that can freely be played by listeners anywhere in the world (in the pre-Internet world, different music was subject to different terms for royalties and use based on the country where the audience was located; having music that is free of geographical restrictions is an important factor for podcaster, since it's virtually impossible to restrict where in the world your work can be played when anyone with an Internet connection could realistically download your show).

Many of these providers also offer music that may be downloaded by listeners an unlimited amount of times (in other words, as a podcaster you would not want to have to be stuck with terms of use that state you only have the rights to distribute a piece of music to 150 listeners). However, before you sign up for any one of these services and start using their music in your podcast episodes, be sure to read all of their FAQ pages, read any contracts or other fine print, and overall be sure you are familiar and comfortable with their terms and conditions. If you can, read other content creators' reviews of that service provider as well, and find out how it has worked out for them.

It is a good idea to set a budget for how much you want to spend on music. A lot of these providers are subscription-based services, which might mean that you have to keep your subscription active in order to continue using the music you downloaded from them in new podcast episodes. This could become expensive over time.

I am not naming any of these providers in this book, not because I don't want to endorse one provider over another, but also because these types of services and providers unfortunately go in and out of business quickly. The information I would provide in this section would likely become out of date quickly. If you want to find a royalty free music provider, a quick search for "royalty free music for podcasters" should provide several results. See if you can find recommendations from other podcasters as well. Take the time to make an informed decision. This might take some time at the start, but will be well worth it once you have settled on a source for your music and have a small library of music saved to your hard drive that you could drop into your shows when editing episodes in the future.

Of course, there is another option. If you are at all musically inclined, you could make your own music for your podcast. This is actually what I did for our intro music. After endless hours spent researching what I could or could not use, I came up with one simple conclusion: I could make my own podcast "theme music." Voila, no worries about copyright law or royalties, because I was the creator and owner of the song!

The ten years I spent in orchestra studying the viola and violin when I was younger finally paid off. Thanks to this music background, I knew how to put together a pretty basic and catchy little intro song for our show. To do this, I again used Garageband, and I put together our podcast music using the free instrument sound bites that came with the application. I spent a few nights building my own catchy 30-second long song, got some feedback on it from others, tweaked it a bit and was done. We then had an intro song for years that I did not need to worry about in terms of copyright or royalties, and it was unique to our show.

I realize that not everyone can make their own intro song, but keep in mind that you could ask around for help. You may have a friend or family member who loves to play around with these kinds of things and might be happy to whip up a little intro song for you. Just be sure that they are composing the music on their own, from their own imagination, and are using resources that are free to use like Garageband's instrument library.

Another great option would be to get a friend or family member who plays in a band to compose an original piece of music for you, record it, and then have them give you their permission to use it as your opening jingle or as bumper music in the podcast. Many content creators are lucky enough to have a close friend or family member who plays an instrument or is in a band. Bands who are looking for extra exposure may have no problem whipping up an original piece (or a few pieces) for you and letting you use it, free of charge, no strings attached, in your podcast. Of course, it would be good form to credit them on your podcast in return. I suggest making sure this kind of arrangement is mutually beneficial: musicians and artists do not always get compensated fairly for the work they do, so I always try to compensate artists as much as I can, if not with money then with something they want in return, such as a credit at the end of each episode or a link or post about the band on your podcast website.

I also highly recommend getting permission in writing from the creator of the music, stating that you (as the podcast producer) have the rights to use the music in your podcast indefinitely, free of charge and free of royalties. In other words, having a brief contract in writing is always a good idea, even if you are working with a friend.

Music can be a sticky, complicated task to find for your podcast. It has become easier in recent years as more and more people become content creators, but no matter what you do, read the fine print. With all of this said, however, unless your podcast is about music, music should not be the main focus of your podcast. I don't even think it is necessary to have intro, outro or bumper music on your podcast if you don't want to include any music. Maybe you would prefer to start your podcast off by talking. Getting right to the point is not always a bad thing. Or, if you do want to use podcast music, you could instead find something short and sweet to use. Whatever you do, music or no music, make sure your choice reflects you and your show.

Since podcasting, by its nature, has an emphasis on audio elements like music, I think that finding the perfect intro song or bumper music for a podcast often consumes too much of the time and focus of podcasters. A lot of podcasters love sound and audio, so naturally, many are also music enthusiasts. As much as I also personally enjoy music, I think the music that is used in your podcast is less important than most other aspects of your show. While it is important to find a musical piece that fits the overall tone and feel you would like to achieve with your podcast as a whole, it is not an area to spend too much time or obsess over when you are first starting out. And as with many other aspects of your podcast, the music you use can organically evolve and change over time if you wish it to.

Cover art and logos

The cover art and the logo for your podcast will be surprisingly important elements of your show. When scrolling through dozens of podcasts on an app, most users will notice a podcast's cover art first. The cover art is their first taste of the podcast, a first impression, and their first exposure to what your podcast might be about. We all know the expression, don't judge a book by its cover. Well, the reality is that most people actually do judge a book by its cover, and they will likely do the same with your podcast.

At the risk of sounding like a Renaissance woman, Jack of

all trades or simply a know it all, I will once again admit that I had some experience with graphic design prior to podcasting. I studied graphic design in school for a while and that helped me tremendously when designing the cover art and logo for our podcast.

You will need to have your cover art and a logo (if you have a logo that is different than your show's cover art) on hand over and over again throughout the years. First of all, you will need it when submitting your podcast to podcast platforms like iTunes or Google Play. We also used our cover art many times as our avatar on various social media profiles. It would also be useful to have the cover art for your show on hand for merch, to send to sponsors, or to send to other podcasts or content creators that may be interested in promoting your show. In short, it will be used a lot through the life of your podcast and will likely be seen by many more people than will ever even listen to your show.

It's also important to have your cover art available in different formats and different sizes. When I first created our podcast cover art, I created it as vector art (in Illustrator) so that it could be resized easily, as needed, without any loss in quality or pixellation. (Vector art can be scaled up or down as needed without any degradation in quality, unlike JPEG or PNG files, often used in Photoshop, which cannot be scaled up without a loss in quality). It's important to be able to scale your cover art both down in size as well as up. Once during our show's run, Apple asked us to upload a larger sized, higher resolution cover art for our podcast because they had changed their standard minimum size requirement for podcast cover art. Because I had made our cover art as a Vector, I could scale it up in size without any change in quality or loss of the integrity of the graphic. Keep this in mind: as technology and industry standard screen resolutions change, you may at some point be asked to upload cover art that is larger than you had originally made when starting the podcast.

If you are not comfortable with designing cover art or working with various sizes and formats of graphics for the web, it is worth having a professional graphic designer (or someone who is familiar with tools like Photoshop or Illustrator) make cover art for you. When you're on a budget, this can be another costly expense, so this is where you will want to be resourceful. Understanding concepts like vector art, resizing an image or graphic without loss of quality, changing color profiles for print vs. screen, using layers, etc., are all important things to already be familiar with when creating podcast art. A friend, family member or colleague who knows about graphic design and can use programs like Illustrator and Photoshop will be helpful.

Having a sense of graphic design and art in general can also be useful for this task. If you have a friend or colleague in marketing or a related field, getting their opinion or advice about your podcast art will be valuable. The cover art for your podcast is important for your show's overall branding, so having someone who is familiar and experienced with branding and design will be very useful.

When it comes to cover art, it is important to do it well from the start. Yes, you can change your cover art at any point later on in your podcast's life, but making a change like that can throw your audience off if they no longer recognize your show's cover art when scrolling through their podcast app. So, doing your podcast's logo, cover art and overall visual branding at the outset is a great idea.

If you are unsure where to begin when it comes to the artwork and overall visual branding of your show, some important things to decide on include:

Colors

Many podcasters use bright colors in their cover art to catch the eye of potential listeners who are browsing through a podcast platform or app and glancing at each show's cover art. I am not sure I recommend this, mainly because it's over used and a lot of podcasts these days have bright, flashy colors in their cover art. When we first started, i read advice that using flashy colors like yellow or bright orange will catch the eye of viewers. Cover art with bright colors do catch the eye, but many podcasters have learned this trick by now. If too many other podcasts' cover art contain bright colours, your cover art will still blend in.

Instead, think about your audience. Are there particular colors that might speak to your work and your audience? For example, for our podcast, we knew that most of our listeners would be women (though not all), and we wanted to evoke a sense of mystery and fairy tale and whimsy. We wanted to stand out, yet also be consistent with the fantasy element evoked by the television series we were talking about. We settled on the use of the color pink in our cover art, which was bright and did catch the eye. However, we used a dark background (that allowed the pink to really pop) with a subtle graphic of a forest to evoke a sense of mystery and fairy tale and whimsy. The cover art we settled on also had a black background so we could easily overlay it on top of any black background (i.e., a black coffee mug, a black t-shirt, a black background in a social media profile, etc.).

We also chose those colors because most of the other podcasts about the same television show at the time had cover art that used blue as their primary color. Our pink, greys and black color scheme was very different than other shows' art about the same topic, and so I felt it would stand out amongst other podcasts on our same topic.

At the risk of falling into stereotypes, we did feel like we would attract a largely female audience and therefore felt pink would catch their eyes in particular versus a mottled blue color scheme. In many ways, we were cheekily using the stereotype of pink being a "girl" color to visually set our podcast cover apart from the other podcasts, most of which were hosted by men, that used blue in their color scheme.

PHOTOGRAPHS

Our podcast cover art consisted simply of text (the name of our show) and a light, vector graphic element (a background that looked like a forest). I was always glad I used vectors rather than a photograph in our cover art, as vectors and other types of simple graphic elements are easy to size up and down. They could also easily be lifted off of the cover art and applied to other graphic elements for use in other images, such as social media headers, banners, and other types of graphics related to the podcast. I used the simple vector tree motif from the background of our cover art in other images and graphic elements, such as the background of our website and banner art in social media profiles.

Some podcasters use photographs in their cover art, but I would hesitate to recommend doing this. Aside from not being able to resize photos well (they cannot be scaled up, only down), photographs usually contain a lot of detail and that level of detail can be difficult for the eye to see when your cover art is sized down and is next to 100s of other podcast cover art in a podcast directory. Keep in mind that your cover art may be displayed anywhere from a very large tv or monitor, to a very tiny icon in the corner of a podcasting app on a phone screen. So, when creating your podcast cover art, study how it looks when it is both big and small in size, as well as how striking or clear it will appear when placed next to other podcast cover art.

STRATEGIC USE OF FONTS

Different fonts can convey different things and send different messages to readers. Our podcast cover art featured two fonts: a light, modern font as well as a cursive font to provide visual contrast.

Serif fonts will subconsciously convey a sense of tradition and formality, whereas sans-serif fonts will appear to be more modern. There are also old fashioned-looking fonts, typewriter fonts, cursive fonts, handwritten fonts... and many more. All of these styles will say different things to the reader. What kind of message do you want to convey to your audience? There probably is a particular font that will help you accomplish that.

There is a whole area of study within graphic design just about fonts, because fonts really do say much more than people realize. For a good introduction on the topic, see the book "Thinking with Type" by Ellen Lupton.

Even the case and alignment of fonts says something. The text in our cover art was lower case. Lower case fonts usually appear more friendly and casual, whereas uppercase fonts will give the sense of being loud, bold, or seem to "shout" at the reader. We also used a cursive element in our fonts, which provided our cover art with a slightly more organic, feminine and casual appearance.

USE OF ARTWORK

Let's talk about our favorite topic some more... copyrights and intellectual rights! As with music, be sure that you have the right to use all elements that appear in your cover art. For this reason, creating a work that is totally original (or hiring someone to do so) is the best option. Make sure you also understand the terms of use of any fonts that you use in the artwork, as some fonts are not available for commercial use without paying for them, with the cost of fonts often running hundreds of dollars. Luckily, dafont and Font Squirrel are two websites with a tremendous amount of unique fonts that are free (!) to download, and there are really clear usage terms for each font detailed on their site. You can even search specifically for fonts that are free for commercial use. (Keep in mind you will probably want to use fonts that are free for commercial use from the outset, in case you eventually end up using your logo in merchandise that you sell.)

Never use photos or graphics you found online unless you have fully read and understood their terms of use. Even though you might see other podcasters using copyrighted graphics or images, that does not mean they should. Or that they will always get away with it. As with music, be careful about reading the fine print on the terms of use and licenses associated with graphics, even ones that you pay for.

Keep it simple

Speaking of which, my graphic design professor used to say that when it comes to graphics, we should always keep it simple. When in doubt, don't overload your cover art with a lot of complicated elements or distracting images. Think about making sure that the viewer's eye will travel to the most important element in the cover art (i.e., the name of your show or a specific icon that represents your show).

More is not always more... a simple, powerful, easy-to-read cover art might feel like playing it safe, but in many cases is also the most effective choice for your podcast cover art. Remember that the ultimate goal with cover art is attracting and enticing new listeners, as well as being instantly recognizable to existing listeners. A simple, visually attractive piece of art that is different than other podcasts that are similar to yours is often the most direct way to accomplish that.

Chapter 11: The Art of Interviews

Live from New York, it's time for you to step up your game as host and interview someone! (Or be interviewed). In your podcast career, you will inevitably have to host a guest on your show or be asked to be a guest on someone else's. The art of interviewing and having an informative, satisfying, entertaining conversation with someone else is an oft-overlooked skill of the podcaster.

We all have watched late-night show hosts who expertly interview guests, one after the next, the flow of conversation punctuated only by commercial breaks. As a viewer, it seems like the host's job is effortless. But when it's your turn to either answer questions in front of an audience - or, harder yet, be the one asking guests questions - you will suddenly realize that interviewing is really an art form. It is certainly more difficult than it looks.

I did not feel like I could write a book about podcasting without including a chapter on how to interview guests. Inevitably, as a podcaster, you will find yourself in the position of having to interview others, whether it's an acquaintance who knows about something related to your show, an expert on a topic that you frequently discuss, a celebrity guest, other podcasters, or even listeners of your show.

The art of interviewing someone is a skill I had a bit of prior experience with before podcasting. When I was a college student, I was interested in going into journalism. Working on my university's student paper and freelancing for a local newspaper for a couple of years gave me a lot of hands-on interviewing experience. That early practice came in handy when interview opportunities came up with our podcast. It was useful to understand the do's and don'ts of interviewing, and how to interview someone in a way that elicits information, seems effortless and facilitates conversation, and avoids mistakes that might result in offending or alienating your guests. There's nothing worse than an interview that is dull to listen to or antagonizes the interviewee and makes them uncomfortable.

BE PREPARED

The first thing I always do before an interview is prepare. As the host, you should maintain control of situations and be ready to handle an interview confidently and competently. To ensure that you are professional and respectful of your guest and their time, do some background research and use that research to pre-plan a few questions. I usually prepare a set of questions before any interview. This will ensure that you have plenty to ask when the interview begins. Be sure to prepare clear, concise and well-researched, relevant questions. There is nothing more cringe-worthy than listening to an interviewer who is caught off guard, who hasn't researched their guest and doesn't know anything about them, and stumbles over what to say or what to ask. You will sound like a rookie if you show up to an interview without having prepared any questions and you may regret, later on, forgetting to ask key things to the person who has so graciously agreed to spend time with you. Not only is it awful to listen to an interview prepared by someone who doesn't know what they are talking about, but it's really rude and perplexing to the guest who showed up expecting to be welcomed and have an intelligent level of conversation commensurate with their expertise on the subject.

Over the years, I have heard a lot of other podcasters say that they want to interview someone in a "casual" way and would like the interview to sound like "two friends talking." Preparing in advance and even pre-planning at least some of the questions you will ask does not mean you will not sound like you are having a casual, friendly conversation. Keeping your environment warm, welcoming and friendly will set this tone. In addition, friends usually know each other quite well. In order to have a friendly conversation with your guest, you should be well-researched about them and understand the topics they will be interested in talking about. Preparing in advance does not preclude a casual, fun conversation with your guest.

Although you will want to pre-plan enough questions to ask your guest, it's also important to remain flexible during the interview. Sometimes, you may really hit it off with your interviewee and go "off script" and want to ask other questions. This is excellent, and will help your conversation seem natural and easy to listen to to audiences. Although, even if you do end up going well off script, I still recommend glancing at your list of questions once in a while during the interview, just to be sure you are hitting on all of the key questions or topics you had hoped to discuss with the individual, time permitting. In other words, remain very flexible and open to allowing the interview to go in unexpected directions, but perhaps jot down some very key themes you really hope to cover or put a couple of questions in bold that you absolutely want to be sure to hit on, if there are any really key items to cover. On the other hand, if the guest goes too far off topic and you feel the the interview is not heading in a productive direction, as the host you have the right to steer them back on topic. You don't need to let the interview spiral out of control if the guest takes the conversation in a direction you don't like. Saying something like, "this is very interesting, but our audience would also like to know more about..." could help direct your guest back on topic. Having your pre-planned questions ready and in front of you during the interview will also help keep your own mind on track and remind you of what you want to cover during the conversation.

Clear, open-ended questions

The very first thing I learned from my editors at the newspaper that I freelanced for was that when interviewing someone, you should always ask open-ended questions. This means you should ask them questions that cannot be answered with a simple "yes" or "no." Nothing stops the flow of conversation faster than a one-word answer by your guest. In other words, instead of asking them,

"Do you like pizza?"

In which case, they may simply say, "Yes!"

You should ask,

"What is your favorite food when you visit New York City?"

This may prompt them to answer with a simple "Pizza," or they may elaborate, but either way, at least they are answering with something that is more interesting than a simple "yes" or "no".

Open-ended questions also lead to more interesting follow-up questions (and, by extension, an interesting conversation). For example, if they answer, "Pizza," you could go on to ask,

"Oh, where is your favourite place to grab pizza?"

"It's in the West Village!"

And suddenly, there's a conversation there. You could go on to chat about what else there is to do in the West Village, some of the top pizza restaurants in that neighbourhood, what the most interesting pizza toppings are, and so forth.

Whereas a simple "yes" or "no" might have resulted in a struggle to grasp a relevant follow-up question, or result in an awkward pause as you gather your thoughts and transition to the next question on your list.

I should mention that although I still used an example of a one-word response ("Pizza"), nine times out of ten a person will respond to an open-ended question with a much longer response, which should make it even easier for you to continue that conversation. Humans love to talk about themselves, and open-ended questions invite them to do this. Interestingly, as a frequent traveler, I also notice this technique used by customs officials: next time you enter the U.S. from a trip abroad, if you are asked a question by a customs officer, it will most likely be open-ended. This is because they know that open-ended questions get travellers to talk and provide more background information with less effort than a yes or no question would.

ALLOW SPACE AND TIME

Common courtesy should tell you that not cutting someone off is the right thing to do when having a conversation with them. The same principle applies for a guest who you are interviewing.

With that said, there are always exceptions. Just pay attention to the context of the discussion: for instance, if several guests are having a rousing discussion and it is appropriate for everyone to talk quickly in a debate-style format, then simply moderating and making sure everyone gets a fair opportunity to speak may be all you can do as host. Conversely, if a guest is struggling to complete their thought or conclude what they are saying, gently stepping in and helping them along to keep the interview paced well may be called for. Being courteous and using your best judgement in the situation should serve you well. In general, though, the idea is to support your guest and be a good host by truly listening to what your guest says.

Sometimes, when interviewing, a host can get nervous and may not pause long enough after a question to allow the guest to respond. This is even easier to forget to do when you are having an audio-only conversation with someone (versus speaking with them face to face or over a video call, where you could usually pick up important visual cues during the conversation). Always remember to wait a few seconds after asking a question to allow your guest to process the question and collect their thoughts before beginning the question. If you continue to talk or add onto the question, it may be hard for your guest to figure out where to respond. In short, allowing pauses between asking questions and having your guest respond, as well as allowing your guest to take short pauses between thoughts when answering your questions, will help them feel like they are being heard. I should note that it seems like pauses are always longer than they really are when you are in the thick of an interview with someone. Later, in your recording, you will usually find that there were fewer long pauses than you thought during the conversation. I'm not sure why this is, other than when we are nervous, we tend to want to fill empty space with conversation, and it is not always necessary!

PUTTING WORDS IN A GUEST'S MOUTH

Although this might seem obvious and like another rule on common courtesy, putting words into someone's mouth is actually a very easy thing for even the most well-meaning of hosts to accidentally do. This was another one of the first rules for interviewing that my mentors taught me when I started working as a freelance journalist, and once I learned how this can very easily happen, suddenly the entire world of journalism changed for me. I observe reporters doing this all of the time in today's mass media, no doubt some intentionally, to manipulate the message of their guests on their shows to further their point. I have also observed other podcasters doing this (I believe unintentionally!) in order to sound friendly or more conversational, but in fact actually stepping on their guests' toes.

Not manipulating someone's message or putting words into their mouths during an interview can be harder to do than you might think.

Essentially, the way this happens (whether unintentionally, as most rookies do, or intentionally, as many irresponsible "journalists" do) is that an interviewer assumes something about the guest and says that assumption out loud or makes that assumption a part of their question for the guest, rather than simply asking the guest point-blank about the topic.

Even in an everyday conversation with friends, you may also be doing this unintentionally. Starting a sentence with something like "You must feel..." is one way it is easy to put words in someone's mouth.

For example, if you are having a conversation with a friend and you are talking about your significant other's new job, your friend might exclaim, "oh, you must be so proud of them!"

The friend clearly means well with this comment, but they are putting words into your mouth by assuming they know how you feel. Instead, the polite thing for the friend to do would be to exclaim how wonderful they think that news is, and then ask you a question about how you feel, such as: "That's wonderful. How do you feel about your husband's new job?" This question still provides you with an opportunity to exclaim how proud you are of your spouse, and this time it was in your own words, not theirs.

In fact, I had this happen to me once when I was the guest on another podcast. What happened was I was being interviewed about my experience as a "female podcaster." Surprisingly enough, being a "female podcaster" was actually a less common thing in 2014, the year that interview took place. The person interviewing me had only ever interviewed one other woman who happened to be a podcaster, compared to the dozens of male podcasters he'd had on his show. (It is an interesting thing that it was so unusual for there to be women who were also podcasters back then, and to be honest I am not sure why podcasting was so unbalanced in 2014, or if the podcaster himself just had no female podcasters in his network. By 2018 or 2019, there are plenty of podcasters who were women, so a conversation like this would have probably seemed bizarre or strange even just a couple of years later.)

We were chatting with the interviewer about our experience as female podcasters, and he then made a series of assumptions about us. Although our exact conversation has been altered here slightly, it went something along the lines of,

"So, you two are one of the few pairs of women podcasters I've heard of. Usually it's a man and a woman, or two men hosting a show," he said.

"Yeah, we are the only two women right now hosting a podcast about Once Upon a Time," one of us mentioned.

"I imagine you must be really proud to be doing something that no one else is doing and representing a different side of the fandom, as two female hosts?" the interviewer said.

"Sure..." we vaguely replied.

In the exchange above, it might seem harmless enough, and it was. The host interviewing us meant well. I am sure he wanted to play up the fact that we were accomplishing something unique or different.

However, what the interviewer did wrong was injecting his own assumptions and opinions about (what he presumed to be) our feelings about our particular experience. By saying, "I imagine you must be..." and following it with a complete guess about our feelings (because how would he know how we feel? He hadn't asked us yet...) he was setting us up to answer the question his way. Although we did not exactly oppose the assumption - again, it was harmless enough and he likely meant well - where he went wrong was that he made the assumption for us rather than just point-blank clearly ask us our feelings on the matter.

An interview should never be about the host. If he had formed a question on the topic without making an assumption first, he would have made it completely about us and enabled us to answer the question in the way that we wanted to, in a way that was closer to our thoughts and truth on the topic. Instead, by framing his assumption in a statement, he was forcing his opinion on the matter on us.

Not only was this making the interview all about him, it

made it harder for us to get our message or feelings across. Sure, we could have disagreed with him totally, but we were trying to be pleasant and the conversation was generally positive, so it felt out of place for us to become too contradictory or negative.

To make us feel more welcome and elicit a more accurate response from us, he should have asked a more neutral open ended question. For instance, the interview should have been more like:

"So, you two are the first pair of co-hosts, who are both women, that I've had on my show. I notice that it's usually a man and a woman, or two men, hosting a podcast," he could have said.

[In a statement, like that, he would have been making it clear this was his assumption.]

"Yeah, we are the only two women right now hosting a podcast about this television show," one of us could have mentioned.

"That's amazing! So how does that impact your podcast?" he might have continued.

"It's not something we think about too much," we would have replied, which would have been true to our experience at the time.

We might have continued, "We believe it is a different dynamic that we are bringing to the table. Being two female hosts differentiates us from other podcasts on the subject, but we believe that every pair of hosts brings their own unique flair to their podcast, regardless of gender. So it's not something that really is a big deal to us."

In other words, he could have still expressed how great he thought it was that we were two women hosts ("That's amazing!"). But after briefly reacting, he once again should have made the question entirely about us, the guests, by asking a neutral, clear, open-ended question ("How does that impact your podcast?"). Note that our answer would not have been the same as how the conversation really went, when his question had been framed by his assumptions. Our answer to a "neutral" version of that question would have been closer to our truth. In other words, if he would have asked us a more direct question, he would have gotten a more honest and clear answer from us.

There are countless ways to manipulate words or to generate negative feelings when you are interviewing others, whether intentionally or unintentionally. To avoid making your guests feel uncomfortable and feel like opinions or answers are being forced on them, as the interviewer, simply remember to keep your opinions and reactions separate from a question. When asking a question, do not use words like "I assume" or "You must" or "I imagine" when talking about the individuals you are interviewing. Let them explain their feelings to you, that's why they are on your show, after all! If you are not sure what their opinion or feelings are, just ask a direct question ("How do you feel about this?" "How has this impacted you?")

Once again, it's important to educate yourself beforehand about your guests. Avoid stereotyping them or bringing your personal biases into the interview. Of course, no one can be perfectly neutral all of the time, nor is anyone expected to be. What you should strive for is to be open-minded, curious and courteous. Making an effort to be informed and mindful can go a long way towards making your guests feel comfortable and to avoid inserting your own personal opinions into what should otherwise be their moment to shine. And of course, if you do make a mistake, a quick apology never hurts.

ADJUSTING YOUR STYLE

The tricky part of interviewing people is that not every single person will react to you, or your questions, in the same way. That is why, while it is important to be prepared with some questions and a general game plan for the interview beforehand, you may have to adjust your style of interview to the person you are with during the actual conversation with them.

When my co-host and I interviewed two of the television series' cast members on stage at a fan convention, we quickly discovered they had very different personalities and very different personas up on stage. One came off as relaxed and bubbly and open to joking around during her interview, with both us and the crowd. To make the conversation with her seem natural, we had to ask questions at a rapid pace and, when appropriate, react with laughter or quick jokes, taking our cues from how she was answering the questions and the dynamic she wished to convey during the conversation.

The other woman we chatted with had a very different persona on stage. She was much more serious and less inclined to connect with us as we asked her questions, focusing more on building a relationship with the crowd in the audience. As a result, we let her answer the questions fully and gave her enough time and space to think and respond between questions. We did not really joke with her on stage and we remained very neutral, allowing her to focus on her connection with the audience, rather than us being a distraction for her. We felt more like facilitators of her discussion up on stage with the audience. In the space of only 30 minutes, we had to completely adjust our interview technique to two different individuals with two very different personalities and styles of connecting with the audience.

That type of impromptu dynamic and the need to think

on your feet and make very fast judgement calls about your guests' style and personality is what I love about interviewing. No two guests are exactly alike. When we were up on stage interviewing the actors, it was thrilling to talk to both of our guests, but also terrifying because anything could have happened.

ADJUSTING TO YOUR GUESTS' LEVEL OF EXPERIENCE

Speaking of adjusting your style to your guest, another thing that you should take into account is how comfortable the person you are interviewing is in front of the microphone or being interviewed. Over the years that we podcasted, we had guests on our show who ranged from former listeners-turned-friends who had never been recorded on anything before, to other podcasters, to experienced actors who had decades of experience being interviewed by the media.

It's important to be mindful that there are differences between interviewing someone who is interviewed often by the media, such as actors or professional experts in a topic, or other podcasters or online influencers, and someone who has never been interviewed or has maybe been on a podcast just once or twice before.

When it comes to professionals who are interviewed, recorded and "perform" for an audience regularly, they are likely going to see an interview on a podcast as just another routine part of their day. It's not something they are likely to think much about until it pops up on their calendar. By the time they speak with you, they've probably done hundreds of interviews over the course of their career and you may not even be the first person interviewing them that day. One of the main things I like to keep in mind when talking with professionals who are used to being interviewed is that they are probably going to see their conversation with you, the podcaster, as little more than a typical media interview or obligation.

With anyone you interview, but especially professionals, you should stick to time limits for your interview. If you spoke with their public relations team, agent, manager or assistant beforehand, you should always provide that person with an idea of how long the interview will last. (We usually asked for 15 minute interviews.) During the interview, I always was careful to stick to the time limits. Having a clock or timer on hand during the interview will help you stick to the time limit (while you are conducting the interview, time will likely fly by, so it's very important to have a clock that will keep track of it). This is the professional thing to do and you will be more likely to be granted more interviews with those individuals in the future (or perhaps others who may be represented by the same agency or manager). It's also courteous.

Another key part of interviewing professionals, and this ties into being courteous about their time, is to keep the interview well-paced and dive right into the topic at hand. There's no need to start with any small talk, other than acknowledging who they are and thanking them for their time. Professionals will be used to just diving into a topic in an interview and so as a host you can get right into it.

However, if you are chatting with someone who has never been interviewed before or has very limited experience being recorded, you will be faced with slightly different challenge. I still suggest sticking to most of the same rules of thumb as professionals, especially with regards to respecting their time. Be clear about how long you want them on your show and stick to it. (They will probably be relieved if you do not go over time!) However, respect that they may be very nervous about coming on the show or speaking on a podcast. I usually gave them a brief overview of what to expect before starting to record. Telling people what to expect helps give them a sense of security and will help calm their nerves, as people are usually most nervous about the unknown.

Next, for someone who is not used to being interviewed, I always started with warm-up or "icebreaker" questions to help them relax into the interview. Asking them an easy question about themselves, such as about their hobbies, pet, how their summer has been so far, can help them ease into the harder questions. Don't linger on these questions for too long, just enough to allow your guest to get used to the format and begin to get used to opening up in front of a microphone

I also suggest deciding for yourself, beforehand, whether you would be willing to edit any part of the interview afterwards. If you have no problem with editing parts of it upon request from your guest, let him or her know that if they really regret saying something, it can be edited afterwards. Knowing that there is the possibility of erasing any big mistakes they might make may help them relax, though be sure to think this through and only say it if you mean it, as it might create more work for you (or whomever edits your podcast).

Lessons learned about interviews

When we first started our podcast, we longed for the day when we could interview someone connected to the actual television series that we were talking about, or thought that having an occasional special guest with us might help attract new listeners to our show.

One of the interesting things we noticed as podcasters was that, contrary to our assumptions on the matter, having special guests or interviewing people, even celebrities from the television show we were podcasting about, did not necessarily generate more listeners or deepen our relationship with our audience.

For one thing, now that I look back at the metrics for the podcast episodes where we had special guests, those were some of our least-listened-to episodes we ever released. This tells me that interviews did not necessarily draw in new or loyal listeners. Usually, when releasing an interview, especially with a celebrity guest, the episode experienced a temporary spike in listeners (especially those who just wanted to listen to the interview with the celebrity). However, that temporary spike never seemed to result in long-term listeners. In the long term, we found that most of our regular listeners did not necessarily choose to listen to those episodes, instead preferring our "normal programming" and normal podcast format.

We liked having special guests host with us or come on to our show because this gave us a new perspective and took us out of our usual routine. However, as I've already touched on in other chapters, we found that our listeners preferred our episodes that had our familiar structure. Judging by how little our episodes with special guests were downloaded, our audience preferred when the two of us hosted our show as per our usual format.

When we first went into the business of podcasting about entertainment, we were in awe of other podcasts that had regular celebrity guests and interviews on their podcasts. When we finally landed a few celebrity guests and went to great pains to have special guest hosts to keep our episodes fresh and new, we were surprised that these were not the episodes our audience gravitated towards. Now that we know this, I wish we had spent less time early on in our podcast worrying about getting guests on our show, and instead enjoying the dynamic we were building as hosts with each other.

Finally, something to keep in mind is that we noticed our dynamic as co-hosts really changed when someone else was added to the mix. With the exception of maybe one or two guests we had on our show, our dynamic with each other significantly changed whenever someone new joined us. Having a third person really disrupted the chemistry we built with each other. In other words, two is company, three's a crowd when it came to us having a special guest host. I suspect this is one of the main reasons most of our regular listeners chose to skip over our "special guest" episodes. Again, the same thing may not happen with you and your podcast, but it is worth keeping in mind that guests can really disrupt the flow and dynamic of your show and may actually turn off, rather than attract, listeners.

Chapter 12: Profiting from your Podcast

Have you ever dreamed of being rich and famous? We probably all have at some point. Although I am not guaranteeing you will ever become rich off of your podcast, it is possible to make a bit of money doing what you love.

So far, I've been avoiding a fairly significant topic in this book: that of fame, glory and wealth.

I realize that you are probably not naive enough to set out to become a podcaster just to become rich and famous. But let's be honest. Every one of us may or may not have had a fleeting fantasy of hitting it big and becoming rich and famous online.

After all, in our present era of glamorous Instagram accounts and YouTube stars, it can feel like every day, an "ordinary" person becomes a big sensation online with hundreds of thousands of followers. A freelance writer starts a podcast that attracts millions of listeners in a month. A couple who enjoys traveling the world puts their adventures on a

YouTube channel and suddenly is taking all-expenses paid luxury vacations and toting around bags full of expensive filming gear. Someone's Tweets go viral and months later, they are elected to public office. Clearly, the social media environment has the potential to make (or break) someone's career. Sometimes, it can even appear to be a fast and easy process.

If you have wild fantasies of hitting it big in the podcast world, or in the social media or online world in general, doing what you love and getting sponsorships and big pay checks for it, then you're not alone. We used to joke about our future "Malibu Podcast House" once we made it big in the podcasting world (knowing full well that having a little podcast about a niche television show wasn't exactly likely to propel us to online stardom). Still, there may have been some small hope from time to time that we might get sponsors, or free merch or, you know, even a second income out of our work.

At the risk of sounding cliche, in the end the real reward of doing the podcast was, well, doing the podcast. We never got paid for it, although we did get a couple of perks. We went to a few events and met a few people, we were given a couple of free tote bags and other small token items from time to time. We had a few sponsorship offers, especially 2 or 3 years into our show, but they were so far off brand from what we were doing (and didn't seem like a good match for our audience) that we turned them down. At the end of the day, we paid for our podcast, and what we got out of it were a lot of laughs, priceless experiences, met some fantastic people, had ups and downs, joys and struggles, and generally speaking, enjoyed a a wonderful, transformative learning experience.

I also believe that doing this podcast had an unexpected benefit, and that was it gave us both something to add to our resume. Creating and hosting a podcast for five years was no small task, and many of our colleagues seemed to find it to be an impressive accomplishment. It was certainly something I added to my resume and listed on my professional website.

Podcasting also allowed us to connect with an industry that we had previously had little exposure to. We now know just a little bit more about the entertainment industry in general, and specifically, broadcast journalism and television and filmmaking. It would be a big stretch to say we were working within those industries, but being podcasters we did get to mingle with some industry professionals. It was both a unique professional experience and provided me with some interesting insights into the career path I had almost chosen long ago when I had considered going into journalism. More broadly speaking, what I got out of this was a new perspective on the business of content creation. Creating content for an audience is hard work, and there's so much more that goes into podcasting than anyone could imagine. It's not difficult for us to see how much work goes into any other kind of content creation. In my opinion, internet celebrities might fall into their role by accident but their sustained success is ultimately a result of a lot of time, effort, hard work and a unique skillset. Understanding more about this set of skills was undoubtedly a valuable experience.

Which probably is not what you want to hear: gaining knowledge and insight is a lot less exciting and romantic of a notion than gaining all-expenses paid trips, thousands of dollars worth of free merch and a full bank account.

Okay, let me back up a second. On the positive side, I actually do think it's possible to create a podcast and expect to make a little money off of it. Even if that money is just going to go back into improving or building upon your podcast, it is not an unrealistic goal to have.

Making a career out of podcasting, on the other hand, will likely take more time, more work, and probably a bit of luck. I would not say it is an impossible goal, but it should not be your main goal, especially when first starting a podcast.

When should you try to make money off of your show?

Your first priority for the first few weeks, even months, of a new podcast should be to create a strong, enticing, enjoyable podcast that people will listen to. Do this and build a strong listener base. Foster a community around your show. Spread the word about your show. At a very basic level, without a podcast that has an audience, you will not really have any chance of profiting off of your work. Focus on strengthening your work first, then worry about profiting off of it second.

There was no set date or time when we felt like we would for sure profit off of our podcast. Instead, it happened organically, just as growth in our listenership happened organically. In the first months of our podcast, listeners were trickling into our show, with listener growth happening slowly but steadily. Still, we never even considered advertising or finding a way to make money, because we felt interest in our podcast remained low after a year. Then, in the hiatus between our first and second seasons of the podcast, we experienced a huge leap in listeners. At the time, we thought maybe it was because we had interviewed one of the television show's main stars over the hiatus (and released a special episode over that break with his interview). While that interview episode had indeed given us a temporary boost in listeners over the break, I believe the real growth over that summer happened because of what I consider, in my mind, to be "The Netflix Effect" (that is my term, by the way). Over our summer hiatus, the most recent Once Upon a Time season that we had just finished podcasting about was released all at once on Netflix. I believe that over that summer, a large number of fans of the show finally caught up on the series. In other words, a large audience who liked the tv series did not watch it live over the fall and winter when it aired on network tv, but rather binged it over the summer on Netflix. When they had finished watching the season on Netflix that summer, I believe they sought out our podcast to have someone to "talk" to about the show. I think that was why we had a large growth in our listeners between our first and second seasons.

All of this to say, every time we saw an increase in the growth of our listeners, it was usually unexpected or felt random and we could not predict it or even really understand it at first (it usually only made sense later, upon reflection). Looking back on our experience, I cannot think of a specific time or a certain threshold when we finally said "oh hey, now we have x number of listeners, let's start advertising." This is something to keep in mind when you think you might want to monetize your podcast: growth with your audience may be slow, may suddenly spike, may level out, etc. In short, there's no guarantee when you'll get enough listeners to feel like you could start to advertise and realistically profit off of advertising. Instead, you may just eventually get a sense of when you might have enough listeners to make pursuing a profit worth your time. The way we knew we might have finally hit a point where it was worth putting time and effort into advertising or profiting off of our show was when we finally got a larger amount of listener engagement with our work; when our listeners were regularly interacting with us on social media, writing emails, and generally when our listenership started to spike around our 2nd and 3rd years.

How to Profit off of Your Podcast

At the time I am writing this, there is still no one cut and dry, uniform way of monetizing a podcast. Unlike very popular video or streaming platforms like YouTube and Twitch, which will insert ads into videos and share some of the advertising revenue with you, no such formula really exists on a mass or universal scale for podcasts. Since you are likely the distributor of your podcast episodes, there is, generally speaking, no magical button to push that will monetize and insert ads into your podcast for you and then pay you every time a listener listens to one of the ads. (There are some exceptions, though, and I'll get to those in a second). This is how, for example, YouTube works: a creator publishes a video, dictates to YouTube whether they want YouTube to insert ads into the video, and then every time someone watches the video, YouTube credits a few cents to their account.

With podcasts, on the other hand, you probably will not have a third party distribution platform to manage the advertising (and revenue) for you. Strictly speaking, in order to have a traditional ad in your show, you will have to insert an ad yourself, either when recording or while editing the episode. That means that you are in charge of the full advertising process yourself, which starts with forging relationships with potential sponsors, putting together agreements or contracts, designing or writing the ad, inserting the ad, invoicing your client for your agreed-upon advertising fee, and so forth. This is a time consuming process and probably the number one reason we never were interested in advertising in our podcast and never ended up having any type of paid advertisements in our show. We both had full time jobs, the podcast was a hobby, so we had zero time to work on being advertising sales and account managers on top of it all. This is key to keep in mind: advertising on your podcast will require a lot of behind-the-scenes work, and you (or someone else) will need to set aside time to manage all of it for your show.

Traditional ads aside, there are a couple of different ways to profit off of your podcast. I am including the most common ways to do so in this chapter, though there are doubtless other ways to make money with your show, and not to mention the fact that tomorrow or next week someone could figure out a newer or better way to do all of this. That's why, as with anything technology-related, you should be open to change and new opportunities. I also recommend using your discernment to decide whether making money off of your podcast is really worth your time and energy, because no matter what route you take, it will consume time and energy and potentially keep you away from doing what you love most, podcasting. You should also consider whether the route you choose will be the right fit for you, your show and your listeners. Once again, return to the concept that you should always keep your audience in mind when making decisions about your show.

POTENTIAL REVENUE OPTION 1: ADVERTISE

As I briefly touched on above, inserting ads into your podcast will probably require you to build relationships with people, companies, or some type of agency. There are a number of platforms out there that claim you can list the name of your podcast on their directory and they will connect you with advertising officials. These are relatively new and didn't exist a few years ago when we started our show, so I have no experience with them, but use common sense when looking into these options. Always take the time to read the fine print, read third-party reviews online of the services if you can, and keep in mind that if it seems to be too good to be true, it probably is.

If you do manage to connect with someone who says they have a service or product they want you to either advertise or talk about on your show, again, I still suggest exercising caution. It may be flattering to be approached and offered an advertising contract. However, on things as sensitive as money and contracts, try to detach yourself from the initial excitement and flattery and be ready to seriously consider the offer and read the fine print.

First of all, ask yourself if advertising is worth it. Again, this will take a lot of time and energy. As essentially hobbyist podcasters, this was where my co-host and I shut down the prospect of any kind of advertising on our podcast: we knew we didn't have the time to manage advertising as carefully as it needed to be.

Second, consider your audience. When you are just starting out or growing your audience, it may be better to go ad-free so people can get into your podcast without the distractions of ads or being taken out of an authentic moment of conversation or interesting material by sponsored messages. In other words, it might come off as a little cheeky or presumptuous if you are advertising on the first episodes of your show. Once your audience knows they like you and your show, then it might be a more appropriate time to work in an ad or two into your show.

If you feel like an offer from an advertiser is good and it is good timing to start to advertise on your show, you should still consider who your audience is and what their interests probably are. Go back to that question we asked earlier: who are they? Is the product or service that you are considering advertising something that would potentially resonate with them or be viewed as valuable or useful for them? If you have a podcast about geology and crystal collecting, inserting an ad about eyeglasses or car oil changes might seem a little off the mark. However, if you have a podcast about geology and crystal collecting and are approached by an online shop selling semiprecious stones and jewelry findings, the type of product that the advertisement would be for might be a more relevant fit and be of interest to your audience.

If you have managed to find a product or service that seems appropriate for your podcast, the next step is to treat the agreement to advertise with the mindfulness it deserves. Ask for a written agreement or contract, and read that document before signing it. Keep in mind you can always redline the terms and conditions of your advertising contract. You are allowed to ask questions, even cross-out or strike out text that makes you uncomfortable, or suggest that things be reworded. A contract or written agreement is, after all, a two-way street, and you have to be happy about what you are signing, the terms they are asking of you, and the amount of money that is being exchanged for what is in essence a service you, as a podcaster, are providing to them.

Once you have carefully reviewed and agreed on the terms and conditions, price and payment terms, it goes without saying that you should always uphold your end of the bargain in the time frame agreed... so that hopefully you will cultivate a good relationship with this new revenue source.

OPTION 2: AFFILIATE LINKS & MARKETING PROGRAMS

Affiliate marketing dates way back: I remember when I first started building websites at the age of 12, and Amazon, which was also in its infancy, was encouraging web designers to put flashy Amazon banners all over their websites in return for a small percentage of the profits of the order that anyone who clicked on the banner and purchased on Amazon.

Affiliate codes are when a company (e.g., Amazon,) provides a content creator with their own unique URL or code that you can then offer to your listeners. Any time a listener uses that URL to shop on a website or purchase an item with your personal link, or if they insert your code into a discount or other type of box during the checkout process when purchasing an item, you will receive a small commission from the sale.

In the past, some content creators have tried to hide that they profit off of these links and codes, but nowadays it's a fairly well-known practice and it is considered good form and courteous to just disclose to your audience upfront that you are giving them an affiliate link or code. (It may even be a legal requirement in some places to disclose affiliate links; be sure you understand your local laws and regulations.) Many followers or listeners do not mind affiliate links, as this is a painless way for them to support a creator they enjoy at no extra charge to them, and it's a way for them to quickly and easily find a link to the product or service being recommended to them. These programs tend to work better when the creators are linking to products that they enjoy, actually use, and that are directly relevant to the discussion at hand. That is why, for instance, on cooking blogs, a blogger may provide an affiliate link to ingredients or kitchen utensils that they used in their recipe. Linking to a special kind of almond milk, mixing bowl and serving dish used and pictured in a post could be perceived as useful and convenient for readers who may want to buy those items as well.

In the years when blogs reigned supreme, bloggers plastered their posts with affiliate links. I have seen blog posts from about 8 or 9 years ago when every other word seemed to contain some sort of affiliate link. Needless to say, using affiliate links or codes is easy on static webpages and blog posts, but they pose a unique challenge for podcasters. After all, your audience isn't reading your work, they are listening. They are not always online when listening to a podcast. They would have to make extra effort to seek out a product mentioned during an episode and be sure to follow your affiliate link or use your code when purchasing it, if they want to support you.

The most common way to deal with this problem is to have a dedicated webpage for your podcast that listeners can go to to find the links and codes to products mentioned. Be sure that listeners can find this page easily and remind them to go to that page when they are ready to purchase the product. Links in social media posts or profiles can also help direct listeners to the right place. Affiliate links can be shared on Twitter, but are more of a challenge on Instagram, as image descriptions on Instagram posts cannot contain links; a link can only be provided in your profile. If you set up an email newsletter for listeners, affiliate links could be inserted into a weekly newsletter.

Last but not least, there are usually terms and conditions - or at the very least, acceptable practices - when it comes to using affiliate links or being a part of an affiliate program. As an example of poor form, I have heard podcasts start their shows with an announcement along the lines of "This podcast is brought to you by___!" Upon further investigation, I often find that the podcasters are being dishonest or misrepresenting their affiliation with that company. The podcasters sometimes invent an "ad" based off of an affiliate code, and not based off of a deeper relationship with that company. In affiliate marketing agreements, usually there is some instruction to affiliates asking them to not create "ads" or make it sound like they have any relationship with the company beyond an affiliate link. Even if there is no strict ban on the practice of faking a sponsorship with the affiliate company, at the very least this is a frowned-upon practice by podcasters and listeners alike because it is deceptive. A podcaster should strive to be honest and genuine, and faking a sponsorship is not that. In short, be courteous to your listeners by being clear that you are providing an affiliate link, and do not pretend to actually be "advertising" for the company itself.

Option 3: listeners as members or patrons

This is rapidly becoming one of the most popular options for content creators to generate revenue. Fans of podcasts are often willing to pay for a "premium" membership to access special or additional content made by that podcast exclusively for their members. The website Patreon has set up an easy platform for content creators like podcasters to build this type of pay-to-play community for fans of their work. Becoming a patron (or rather, "patreon") of a podcast allows listeners to feel like they are directly helping out the creators of the show, without being forced to sit through ads or remember to use affiliate links. Setting up a subscription option, or creating a Patreon page, gives listeners who really love your show and want to give back to you an option to pay a set amount of money (anything from $1 to $100 a month, or more) as a sponsor of your work. In return for their sponsorship, many podcasters will provide perks like access to a patron-only blog where members can listen to extra podcast content or have access to special episodes or live streams. Podcasters also tend to give members greater access to interactions with them through a messaging system or participation in a secret online group, like a members-only Facebook group.

The benefits of this option are that sometimes people just want to help podcasters they love out and feel they owe podcasters something for all of the hard work they do on their show. The idea is not so different than donating to a cause you are passionate about, or to a public broadcasting station. Setting up a way for people to donate to you provides them an outlet to give you a donation if they feel compelled to.

Furthermore, having members-only groups for the biggest fans of your podcast to interact in can sometimes result in stronger and more bonded communities than public groups for your podcast will provide. Fans of a show might not want to post a public comment that could be viewed by anyone, and therefore prefer the more protected setting of a private group. With sites like Facebook often sharing or making comments in public groups available to their other friends and members of their network without their consent, interacting in a private group is appealing for those who do not wish to have such public conversations on a topic they enjoy. Requiring fans to pay for their membership in secret groups results in a barrier to entry and a smaller, more intimate group of members.

Option 4: Sell your own products or services

If you are starting a podcast related to your field or industry and you also happen to be a business owner, entrepreneur or freelancer, it may be worth talking about the "real life" services or products you provide, and where to find them. As always, use discretion, as you don't want to constantly spam your listeners with promotion for yourself or business, but working in a mention of your product or service from time to time might help you raise awareness for your business and drive traffic to your other products and services.

Of course, this is best done if you are mentioning services that you personally provide or products that you personally, as an individual, sell. Mentioning your employer or their products without their permission would not be wise (for privacy reasons for yourself and for them, among other factors). Most medium to large companies have a "social media policy" that essentially states you are not allowed to represent your employer on your own social media accounts (which would include a podcast) without their permission.

You might also decide to set up an online shop on an e-commerce website like Etsy while you are podcasting and stock it with items related to your podcast (or even items that could pique the interest of your audience). For instance, I listen to a podcast hosted by someone who has an Etsy shop. She mentions it from time to time on her show, and is clear that her shop is her main source of income. The implication is that by checking out her Etsy site and possibly buying an item, listeners are continuing to support her work and her podcast.

There are also many print-on-demand websites that will allow you to set up your own shop. This gives content creators the opportunity to create merchandise for their podcast, such as t-shirts or coffee mugs, printed with a logo or cover art of their podcast, or using a funny saying or graphic that is related to a running joke or a reoccurring theme in their podcast. My co-host and I were surprised to receive a number of requests for merchandise like this when we were podcasting, though unfortunately we never had the time required to set up the infrastructure for an online shop. (Our listeners liked a coffee mug that my co-host had special-made as a gift for me with our show's logo on it, and many of our listeners wanted to buy the mug). For a podcast with a lot of listeners, a shop with some items related to your show could be a fun and viable way to make a little extra money.

Option 5: Promote yourself

When it comes down to it, monetizing a podcast can be complicated and time-consuming. It will take away from the time you could otherwise use to work on improving the podcast itself. If there are several people working on your podcast, it might be wise to assign the revenue generation duties of the podcast to one person, so that not everyone involved gets bogged down in it.

If you decide to take on the costs of building and maintaining a podcast yourself, I believe that podcasting can lead to some professional benefits down the road that cannot be bought with money. I found that, as a creative type who has always worked in marketing or sales-related fields, actually creating and sustaining a podcast became something that many of my colleagues found impressive and memorable. Being a podcaster had an impact on my professional life and elevated my professional profile. Some of my colleagues even ended up helping me with the podcast by connecting me with others who they knew in the industry. In this way, I will likely continue to "profit" off of my experience as a podcaster for many years to come. Podcasting was about upgrading my professional skills, and that experience could still lead to more professional opportunities for me in the future that I would not have ever had without being a podcaster. You cannot put a price tag on that.

Chapter 13: Podcasting your Reality

My final parting thoughts to you as you set off on your podcasting adventure.

As someone who has loved writing and art her entire life, I don't think there's a more noble pursuit in life than creating things. Whether I am creating podcasts, art, books, gardens, food... I know that I am bringing something to this world. Something that I designed to give others joy, satisfaction or happiness.

Perhaps the most important lesson I took away from having a podcast for four years was that podcasting is not something you think about doing. A podcast is not something you say that you would like to have. Podcasting is not a thing you want to do.

To be a podcaster, you have to just do the podcast. It's as simple - and as complex - as that.

I frequently go to yoga classes, and quite often my yoga teachers will start a yoga class by telling us "thank yourself for making it to class today." And yes, getting to class in the first place can sometimes feel like a bigger hurdle than actually completing the one hour yoga class. Likewise, as with any creative pursuit, the biggest battle of making a podcast is, well, actually getting started and doing the podcast. That's because it's easy to talk about podcasting. It's easy to purchase and read this book, buy the equipment, and set up a space for podcasting.

But if you want to start a podcast, you have to show up at the microphone and record.

As I mentioned earlier in this book, I believe that everyone has something important and significant to share with the world. Podcasting is a fantastic medium for conveying your personal truth and the message you wish to contribute. Having a podcast is also amazing for forging connections with others who share your passion. By creating a podcast, you will be creating a path through which you can share your story, teach others, and connect with a community.

These reasons - and more - are why it is worth getting started and actually doing the podcast. Always remember why you wanted to start a podcast, and use that to motivate you as you continue forth with the project.

Before I leave you to begin your own podcast, I thought I would share a few final, parting thoughts that I believe would have helped me if someone had told me these before we started our podcast.

Parting Thought 1: Podcasting is a lifestyle

To quote a wise Jedi master, "Do, or do not. There is no try." There is no podcast from an idea, or from the purchase of a microphone or how-to book. There is only a podcast if you actually start podcasting.

To get started (and to move past that significant first hurdle of actually starting), mark down the release date of your first episode in a calendar. Then write down your second, then your third episodes. And stick to the schedule you have created.

Aside from setting a firm date to release your first episode, as I also touched upon in the chapter on consistency, being consistent with a recording and release schedule is beneficial to both yourself and your listeners.

By working an episode release schedule into your weekly or monthly routine, you will begin to schedule your life around the podcast, making space for it, scheduling in work like researching, recording, editing, releasing or promoting each episode. Even though I am, by nature, not a huge fan of setting a rigid schedule, with our podcast I found it useful to have a set schedule because that was how I made sure I always had time to focus on the podcast during my busy week. By creating a steady habit of podcasting on certain days and releasing new episodes on specific days and times, I found myself getting into a podcasting routine. Once I was in the routine, I didn't want to get out of it. I became used to it, just like one might grow used to a weekly workout class or reoccurring monthly meeting at work.

By knowing I had to podcast on Monday nights and edit and release the episodes by Tuesday morning, and that this was a non-negotiable schedule, we always got our podcast done and released on time.

THOUGHT 2: KEEP GOING!

It's not easy to keep podcasting, even on a fairly rigid or regular schedule. When, years ago, a fellow podcaster mentioned to us that most podcasts only survive, on average, 7 episodes before their creators give up and quit podcasting, I remember thinking that it was such a shame to have put all of that hard work into those seven episodes, begin to build an audience, only to have it all end prematurely. Early on in podcasting - even during the first few dozen or so episodes - audiences may not be very large, they might not be writing in or interacting with you, and the tangible rewards of podcasting may be few or nonexistent.

This can be difficult to handle because we live in a culture of instant gratification. When I post a new image on Instagram, I hope to get a certain number of likes within the first few minutes of my post. If I don't, I grow concerned, even worried that my post that day didn't live up to expectations or that I am doing something wrong. In other words, I, like so many others, have been trained to expect instant success and instant reward for whatever I do. In many ways, our world is manufactured this way: Instagram has taught us to expect and enjoy that instant approval from our peers if we did "well" on the task at hand, rewarding us with dozens of likes and a boost in the visibility of our post or page thanks to their algorithms.

However, expecting instant gratification on the scale of hundreds or thousands of "likes" from listeners right out of the gate is not the way to start out as a podcaster. That isn't to say you won't get a few early rewards - a positive comment from a new listener, a like on your podcast's social media page - but the key is to be satisfied with those small successes rather than expect or fixate on big, dramatic results. In other words, forget what Instagram and other social media sites have taught you. Podcasting is a different medium, and if you don't get the results (aka., traffic, downloads, number of listeners, and so on,) you had hoped for immediately (or ever), that's fine and totally normal.

With some exceptions (I mean, I would be remiss not to

acknowledge that some people's podcasts do take off really fast or an early podcast episode could go viral), a typical podcast will take a while to reveal tangible results. In fact, sometimes results, such as growth in listenership or the increase in comments and other types of engagement by your listeners, may trickle in so slowly that you will barely notice it. And then it will gradually increase over time. The amount of listeners we had increased at a snail's pace in the first months and even years of our podcast; if we had only used the number of listeners of our podcast to gauge our success, we would have grown frustrated quickly and probably would have given up very early on.

In the early days of your podcasting career, really one of the only things you can do is just keep going. Podcasting became such a routine part of my life, it was just like going to work every day or taking out the recycling on Tuesdays. (Except it was a little bit more fun than those chores!) And you know what? Thanks to simply sticking with it, the result was that we did not give up after 7 episodes. We released 30 or so episodes our first year and continued this same, consistent schedule for over four years.

THOUGHT 3: CELEBRATE THE SMALL STUFF

While I had ambitions and aspirations for our podcast, I did not hold myself to any hard and fast goals. I tried to not compare our podcast to others. This was sometimes hard. I remember when we first started podcasting, other fan podcasters about the same television show received screener copies of the one of the season premiers of the show, and we did not. There were other little indicators of success like that which we fell short on, which bothered me.

To avoid growing unnecessarily frustrated, we never set any specific measurements or goals that we had to reach along the lines of "we must have x number of listeners by y date" or "we need to get screeners next year." Instead, my co-host and I promised each other that we would have fun with the podcast, and that if only a dozen people listened to it, we would be satisfied.

Because of this, I didn't look too closely at how many listeners we had (or didn't have) in the first year or two of podcasting. The numbers were only in the hundreds, which, in an era of social media influencers who get 20,000 likes on a post in only a few hours, easily could have seemed discouraging. (Looking back at it now, I think that 100 or so dedicated listeners week to week - which amounts to a small theater filled with people - actually is nothing to scoff at, but it's all relative, isn't it?) Anyways, instead of fixating on the actual number of subscribers who were listening to our podcast week to week, I started to count it as a success if I saw our listenership steadily increasing over several months. I remember one year after we started podcasting, I looked at our stats from that year and noticed that there had been an exponential increase in subscribers and downloads over the course of just 12 months. We had started our podcasting journey with about a dozen or so listeners in its first few weeks, and ended the year with hundreds of subscribers. When I looked at a graph showing episode downloads over the course of that year, the graph reflected steady, steep listenership growth. That upward growth was really satisfying and motivating. But I had to wait a year to notice that, as the growth was only noteworthy after that length of time.

Another measure of success that we celebrated was when we started to get regular feedback and comments from listeners. It was motivating when we received regular notes, emails and Tweets from listeners. They sent us encouraging words and shared their reactions to things we had discussed on the podcast. To me, it was exciting that our listeners cared enough to take a few minutes to respond to our work. These interactions were an early token of our success. Plus, it was nice to hear from listeners because it assured me that there were, in fact, other human beings actually listening to and appreciating the work we were doing.

Thought 4: evolutive thinking

In order to not get overly bored with our podcasting routine, I took time every so often to pause and think about what I enjoyed most about our podcast. If you are podcasting with someone else (or a few others), check in with each other regularly and have a meta conversation about your podcast. Communicate with each other about what you are enjoying most (or least) about the show, or the process of podcasting in general, and don't be afraid to discuss whether redistributing the workload in some way might be helpful. Adjusting your methods and rearranging your routine according to what everyone likes and dislikes from time to time could help keep you and your podcast partners motivated.

For example, if one of you enjoys maintaining the social media accounts and interactions for your podcast and another one of you enjoys editing the podcast episodes, then distribute the work in such a way that whoever enjoys editing does more editing and whoever enjoys managing the social media is doing the digital marketing side of the show. Likewise, if any of you grow bored with certain tasks, just because early on you decided one of you would do more editing and another one of you would do more pre-show planning and research, doesn't mean you have to always stick to those roles throughout the life of your podcast. Don't be afraid to switch it up, learn new things, and trade roles behind the scenes, so that everyone will be less likely to get bored with the tasks. Rotating jobs also helps everyone have more opportunities to learn new things and evolve their individual roles over time. If you are a solo podcaster and creating your podcast on your own, and find certain aspects of your show tedious, consider getting outside help or even seeing if you could hire someone to work with you every so often.

Thought 5: practice podcaster self-care

"Self-care" is a bit of an overused term, but the concept of stepping back and telling yourself that you don't need to perform at 100% every second of the day is an important thing to do. Self care is about slowing down once in a while so that you can recharge your batteries and return stronger than ever afterwards.

Unless you are being paid to podcast as a part of your job (and if so, a pay check and continued employment is certainly a powerful motivator, but also brings with it many different types of challenges!), podcasting may become draining over time, no matter how much you are enjoying the process and how rewarding you find podcasting to be. Like any hobby, it can have its downsides every once in a while. Too much of one thing (even too much of a good thing), can ultimately be draining and stressful. As much as I enjoyed podcasting, every couple of weeks I needed a break from it, as the pressure of getting an episode out on time and the constant worry about negative comments and feedback often lingered in the back of my mind.

The breaks we took between podcast seasons were always welcome. We usually had to put out podcast episodes on a weekly basis in the fall and early winter, from September to December. We then usually got January and February off, before putting out podcast episodes from March through May. And then we had downtime again from June through

August. Although sometimes the 10 to 12-week stretches of podcasting on a weekly basis in the fall and spring got to be very tiring, at least we always had a 2 to 3 month break to look forward to at the end of those stretches. By the end of 12 weeks of podcasting on a weekly basis, I was more than ready for some down time and for some weeks when I didn't have to podcast every Monday night.

If you hit a point when you have started your podcast strong and have been putting out new episodes for a few weeks, but are suddenly desperate for a break, then take one! Even thought the day to day and the nitty-gritty of podcasting isn't always fun, in general podcasting should be enjoyable and something that you can, in the larger picture, feel very proud of and accomplished to have done. If you ever start to feel like this isn't happening, take a step back, take a break, and after a while, think about what you can do to enjoy the process more upon your return.

REALLY FINAL THOUGHTS

As I was writing this book, I began to realize that there really is no rulebook or instruction manual for making a podcast. I hope that my thoughts, anecdotes, experiences and reflections have provided some insight into the approaches you might (or might not) take with your own show. I set out to provide insight, inspiration and a kind of handbook to help you with podcasting. I also hope that you now see that you can do it all on a budget that is reasonable to you.

As with any creative endeavour, however, there is no one set of rules that you have to follow to get a podcast done, nor is there an instruction manual for how to enjoy success with your work. In the end, I recommend being true to yourself and what you want to accomplish and who you hope to connect with. I also suggest that you pace yourself. Although podcasting requires a degree of discipline and a routine, it also benefits from you being able to relax and enjoy yourself throughout the process.

Regardless of whether you have started a podcast, will start a podcast, succeed, or not-quite-succeed-but-grow from the experience, what I wish most of all is for you to learn something along the way and hopefully find enjoyment, insight and human connection.

The art of podcasting is like any other art. At the end of the day, it's about creating what you can and sharing your truth with the world. That is its heart and soul. The microphones, the websites and servers, are all just the tools of the trade. As the painter Frida Kahlo once said about her work, "I don't paint dreams or nightmares, I paint my own reality."

How will you paint your reality with your podcast?

Acknowledgements

Thank you for being a part of my podcast journey by reading this book, and I wish you much success along your own podcasting adventures.

Many thanks to my podcasting co-host and partner, Brittany, for agreeing to go on a podcasting adventure with me back in 2013 and seeing our beloved podcast through to the end in 2018. I also appreciate your support this year as I wrote this book, as well as your contribution to it in the beginning. Cheers to a wonderful, enlightening, creative, and fun journey.

Thank you to my mom, Jane, one of our podcast's first listeners and for proofreading support on this book.

Thank you to my husband, Alex, for your enthusiasm for the podcast and encouragement when I told you I was writing this book.

Many thanks to our Once Upon a Podcast audience who stuck with us through thick and thin and taught me that the most precious part of being a podcaster was being a part of a community. I love that I got to know so many of you! You taught me so much and I hope to meet many more of you in the years to come.

And last but not least, my tremendous gratitude to my family members, colleagues, and fellow podcasters who provided support along the way.

Sources consulted when writing this book:

RSS, a Brief Introduction - https://www.ncbi.nlm.nih.gov/pmc/articles/PMC2565593/

"Podcast Listenership" (Pew Research Center) - https://www.journalism.org/chart/sotnm-radio-podcast-listening/

"Podcast" (Wikipedia) - https://en.wikipedia.org/wiki/Podcast

About the Author

Amanda Greenman began studying broadcast journalism in high school and has been dabbling in media, podcasting and writing ever since. Now a writer and editor, she has a Bachelor of Arts in Art History & Communication Studies from McGill University and a Master of Adult Education from St. Francis Xavier University. Born in Chicago and raised in the Great Lakes State, she enjoys travel and lives with her husband and two spirited guinea pigs.

You can visit her at www.amandagreenman.com

www.ingramcontent.com/pod-product-compliance
Lightning Source LLC
Chambersburg PA
CBHW051233050326
40689CB00007B/909